Molecular Biology of Prostate Cancer

Molecular Biology of Prostate Cancer

Editors
M. Wirth · J. E. Altwein · B. Schmitz-Dräger · S. Kuptz

Walter de Gruyter
Berlin · New York 1998

Editors

Prof. Dr. M. Wirth
Universitätsklinikum Dresden
Klinik für Urologie
Fetscherstr. 74
01307 Dresden
Germany

Prof. Dr. B. Schmitz-Dräger
Heinrich-Heine-Universität Düsseldorf
Klinik für Urologie
Moorenstr. 5
40225 Düsseldorf
Germany

Prof. Dr. J. E. Altwein
Krankenhaus der Barmherzigen Brüder
Klinik für Urologie
Romanstr. 93
80639 München
Germany

Dr. S. Kuptz
Zeneca GmbH
Forschung Onkologie/Endokrinologie
Otto-Hahn-Straße
68723 Plankstadt
Germany

Die Deutsche Bibliothek – Cataloging-in-Publication-Data

Molecular biology of prostate cancer / ed. M. Wirth ... · Berlin ; New York : de Gruyter, 1998
ISBN 3-11-016159-1

Reproductions, typesetting and printing: Arthur Collignon GmbH, Berlin – Binding: Lüderitz & Bauer GmbH, Berlin – Cover design: Rudolf Hübler, Berlin
Printed in Germany

Preface

Prostate cancer is one of the most common malignant diseases affecting men in the western world. With an estimated number of 334 500 new cases in 1997 in USA prostate cancer has displaced lung cancer from first place in the table of annual incidence rates in men. The enormous significance which prostate cancer has acquired is reflected in the increasing number of scientific publications.

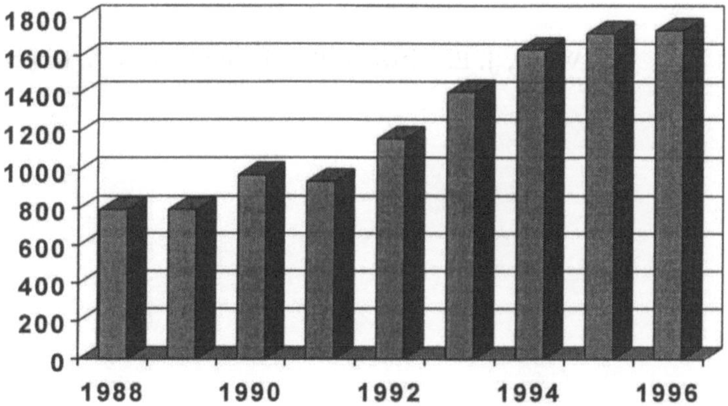

Fig. 1 Number of publications on prostate cancer from 1988 to 1996.

Our knowledge however on the aetiology, the natural progress and the biological potential of this tumor is insufficient. Even now the majority of patients have already reached an advanced stage of the disease when it is diagnosed. In this stage prostate cancer is principally incurable.

Molecular biology is one field of research which has developed very rapidly in recent years. The knowledge gained on the function of tumor genes and tumor proteins has contributed considerably to the understanding of molecular and cell biology basis of malignant growths. Based on these principles new approaches to prevention, diagnosis and treatment of tumors are being developed. This however requires close co-operation between the doctor and the pure researcher.

The challenges presented in the clinical urology and basic science of prostate cancer make it necessary to follow new paths. Our idea was to organise a symposium in which scientists and medical doctors introduce the newest biomolecular knowledge in prostate cancer and discuss its significance for clinical use. To structure the abundance of information we have divided the symposium into four sections. Session 1 was concerned with the molecular principles of the genesis of prostate cancer, the involvement of oncogenes and tumor suppressor genes. Session II involved the changes of cell-cell contacts, the defects in androgen receptors and their effect on treatment with antiandrogens. Session III explained drug resistance mechanisms and new therapeutic principles such as antiangiogenesis inhibitors, intermittent androgen ablation and gene therapy. Session IV was concerned with the diagnosis of prostate cancer, new procedures such as RT-PCR and possible new markers.

All contributions have been included in this book. They provide the reader with the opportunity to consult the wealth of information while creating a basis for new ideas and approaches in the research and treatment of prostate cancer. Only so is it possible to discover new paths for the management of prostate cancer. This symposium has been designed to provide the basis to further stimulate this evolution.

M. Wirth, J. E. Altwein, B. Schmitz-Dräger, S. Kuptz

Contents

List of first-mentioned contributors

Prof. Dr. J. E. Altwein
Krankenhaus der Barmherzigen Brüder
Abteilung für Urologie
Romanstr. 93
80639 München
Germany

Dr. M. C. Beduschi
The University of Michigan
1500 East Medical Center Drive
Ann Arbor, Michigan 48109
USA

Dr. M. J. G. Bussemakers
University Hospital Nijmegen
Urology Research Laboratory (814 URL)
PO Box 9101
6500 HB Nijmegen
The Netherlands

Prof. Dr. S. F. Brewster
Churchill Hospital
Department of Urology
Oxford OX3 7LJ
U. K.

Dr. L. W. K. Chung
University of Virginia
Health Science Center
Department of Urology
Medical Center Box 422
Charlottesville, VA 22908
USA

Dr. W. J. Ellis
University of Washington
School of Medicine
Department of Urology
Box 356510
Seattle, WA 98195–6510
USA

Prof. Dr. B. Furr
Zeneca Pharmaceuticals
Mereside, Alderly Park
Macclesfield
Cheshire, SK 10 4TG
U. K.

Dr. M. Gleave
The University of British Columbia
Division of Urology, UBC
910 West 10th Avenue
Vancouver, B. C. V5Z 4E3
Canada

Prof. Dr. K. Griffiths
University of Wales
College of Medicine
Tenovus Cancer Research Centre
Tenovus Building, Heath Park
Cardiff CF4 4XX
U. K.

PD Dr. R. Heicapell
Universitätsklinikum Benjamin Franklin
Urologische Klinik u. Poliklinik
Hindenburgdamm 30
12200 Berlin
Germany

Univ. Doz. Dr. H. Klocker
Universitätsklinik für Urologie
Anichstr. 35
6020 Innsbruck
Austria

PD Dr. Th. Licht
National Cancer Institute, DBS, NIH
Laboratory of Molecular Biology
Building 37, Room 1B23
37 Convent Drive, MSC 4255
Bethesda, MD 20892–4255
USA

Prof. Dr. D. Marmé
Universität Freiburg
Institut für Molekularmedizin
Breisacher Str. 117 D
79106 Freiburg
Germany

Prof. Dr. C. Olsson
Squier Urological Clinic
The Presbyterian Hospital
161 Fort Washington Avenue
New York, NY 10032
USA

Prof. Dr. R. Paul
Urologische Klinik u. Poliklinik
der TU München
Klinikum Rechts der Isar
Ismaninger Str. 22
81675 München
Germany

Prof. Dr. M. Peehl
Stanford University
Department of Urology
School of Medicine
Stanford, CA 94305−5118
USA

Dr. Ch. P. Pilarsky
MetaGen GmbH
Ihnestr. 63
14195 Berlin
Germany

PD Dr. G. Sauter
Kantonsspital Basel Universitätskliniken
Institut für Pathologie
Schönbeinstr. 40
4003 Basel
Switzerland

Dr. D. H. J. Schamhart
University of Amsterdam
Department of Urology
Meibergdreef 9
1105 AZ Amsterdam
The Netherlands

Dr. Bettina Schmidt
Heinrich-Heine-Universität
Klinik für Urologie
Moorenstr. 5
40225 Düsseldorf
Germany

Prof. Dr. M. J. Siegsmund
Universität Heidelberg
Fakultät für Klinische Medizin
Urologische Klinik d. Stadt Mannheim
Theodor-Kutzer-Ufer 1−3
68135 Mannheim
Germany

Cytogenetic alterations in prostate cancer

G. Sauter

1. Introduction

Development and progression of tumors is driven by an accumulation of genomic alterations. These genetic changes usually result in a disturbed function of genes involved in the control of key mechanisms for cellular growth and differentiation such as proliferation, angiogenesis, cell-cell and cell-stroma adhesion, apoptosis, DNA mismatch repair, or preservation of genomic stability. The genes involved in tumorigenesis can be categorized as ***tumor suppressor genes*** (tumorigenic through inactivation) or *oncogenes* (tumorigenic through overexpression). Cytogenetic analyses are relevant for the examination of tumor suppressor genes and oncogenes since alterations of these genes are frequently associated with quantitative or structural chromosomal alterations. A tumor suppressor gene inactivation is frequently associated with a deletion of one allele and an oncogene overexpression can be due to a gene copy number increase (gene amplification) or structural chromosomal changes (translocation, inversion).

The human genome comprises about 100 000 genes, only a small fraction of which are currently known. Given the high proportion of unknown genes, many oncogenes and tumor suppressor genes may be yet unidentified. Considering the strong association of chromosomal deletions with tumor suppressor gene inactivation and the relationship between DNA sequence amplification and oncogene overexpression, it is postulated that the detection of frequently deleted or amplified areas in the genome may pinpoint the site of unidentified oncogenes and tumor suppressor genes. The evaluation of cytogenetic changes is therefore an important starting point for the identification of novel tumor-associated genes.

Conventional cytogenetic analysis always includes a short-term culturing of tumor cells which are subsequently arrested in metaphase or prometaphase. After dropping these cells on glass slides, the chromosomes are spread and can be analyzed after Giemsa staining. Major limitations of classical cytogenetics are the need for fresh tissue and the risk to select non-representative clones (neoplastic or non-neoplastic) during cell culturing. The simultaneous analysis of the entire genome and the ability to detect structural changes (translocations, inversions) are the strongest advantages of cytogenetics. It is likely that modern hybridization procedures allowing the simultaneous identification of all chromosomes in different colors will further improve the power of cytogenetics [32]. Molecular cytogenetic methods including fluorescence in situ hybridization (FISH) and comparative genomic hybridization (CGH) have the potential to substantially improve our understanding of the genetic changes in prostate cancer. FISH allows visualization (and therefore quantitation) of chromosomes and genes on a cell by cell basis [16, 21]. CGH is based on the simultaneous hybridization of differentially labeled tumor and normal DNA's to normal metaphase chromosomes. CGH allows the detection of all amplifications and deletions (> 10 megabases) in a single examination [19]. Since archival tissue can be analyzed by both FISH and CGH, these techniques can be used to examine tumor sets with clinical follow up information to evaluate the clinical significance of chromosomal aberrations.

This review summarizes the typical cytogenetic and molecular cytogenetic findings in prostate cancer. Emphasis is placed on these studies investigating the clinical significance of cytogenetic analyses.

2. Cytogenetic changes in prostate cancer

Genetic changes underlying initiation and progression of prostate cancer are poorly known. Classical cytogenetic studies are difficult to carry out because of the preferential growth of non-malignant cells. Therefore cytogenetic analysis is often impossible in prostate cancer. Among more than 6000 solid tumors examined by cytogenetics by the end of 1994 there were only about 50 prostatic carcinoms showing cytogenetic aberrations. Despite these problems cytogenetic analyses have identified several alterations as being frequent in prostate cancer such as gains of chromosome 7, losses of the Y chromosome and regional deletions at 7q, 8p, 10q, and 16q.

2.1 Y chromosome loss

Y chromosome losses are frequent in prostate cancer [6, 22]. Using FISH, Van Dekken et al. [37] could show that Y chromosome losses in prostate carcinoms occur in epithelial but not in stromal cells. König et al. [22] found a Y loss in 53% of their 42 carcinomas of the prostate. A Y loss was significantly more frequent in prostate carcinomas with unfavorable outcome than in less malignant tumors in another FISH study [1]. Despite the high frequency of Y losses in prostate cancer and many other tumors [26], the biological significance of Y losses in tumors has remained unclear. Studies describing Y losses as sole abnormalities in various tumors including prostate carcinomas [26] are arguments for a relevant role of the Y chromosome, which has been suggested as the site of a potential master anti-oncogene [27]. Other studies have found Y losses in various non-neoplastic tissues [7, 13, 31, 36]. It has been shown that prevalence of Y chromosome losses increases with age in normal bone marrow cells [36]. Y losses have been found by FISH in a relevant fraction of cells (> 20%) in 8 of 28 benign prostatic hyperplasias [3]. These findings seem to suggest that Y losses may be a random, clinically irrelevant phenomenon in prostatic tissue [7, 13, 36]. While it can not entirely be excluded that Y losses can contribute to malignant transformation of prostatic epithelial cells, the findings in non-neoplastic tissues show clearly that a Y chromosome loss can not be used as an indicator of malignancy in prostatic tissue.

2.2 Chromosome 7 alterations

There are several lines of evidence that genomic changes involving chromosome 7 are relevant for prostate cancer biology. Typical chromosome 7 changes found in prostate cancer include polysomy 7 and 7q deletions [26].

Early cytogenetic studies have shown that chromosome 7 overrepresentation (trisomy 7) is frequently found in prostate cancer [26]. These results have later been confirmed by CGH and FISH studies [1, 2, 6, 10, 39]. Fig. 1 shows cells of a prostatic cancer with a chromosome 7 polysomy.

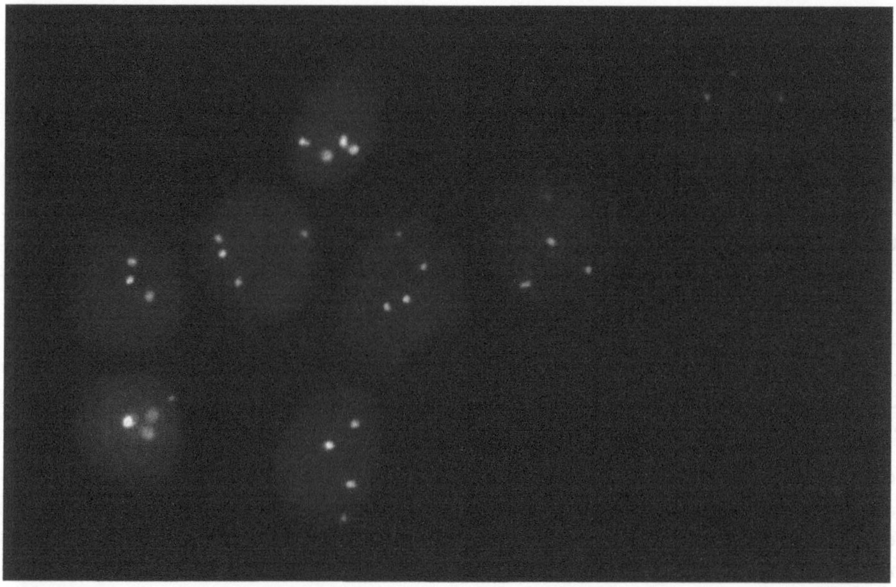

Fig. 1 Chromosome 7 polysomy. In situ hybridization with a centromere 7 probe and DAPI counterstaining. Group of prostate cancer cells containing 4 chromosome 7 signals per cell (tetrasomy). Magnification 600x.

Several reports have provided evidence for a role of a polysomy 7 for prostate cancer progression [1, 5]. The strongest data in this regard have been provided by Alcaraz et al. [1] showing a polysomy 7 in 24 of 25 tumors of patients dying within 3 years after radical prostatectomy but in only 3 of 25 tumors of patients surviving more than 5 years ($p < 0.0001$). While these data are promising it must not be forgotten that polysomies of chromosomes detectable by FISH are frequently the result of an event involving more or less the entire genome (DNA aneuploidy). FISH analyses of multiple chromosomes in individual tumors have shown that polysomies are usually not restricted to one chromosome but involve several, if not all, chromosomes examined in most tumors [1, 30]. It is therefore not surprising that polysomies of all 12 chromosomes examined besides chromosome 7 by Alcaraz et al. [1] were linked to poor patient prognosis although to a lesser extent than polysomy 7. Further studies are needed to determine whether polysomy 7 is of particular clinical relevance for prostate cancer. It is important to note that a chromosome 7 gain is another cytogenetic alteration that has been frequently found in various non-neoplastic conditions [17]. Using FISH, Aly et al. [3] found a polysomy 7 (in > 10% of cells) in 7 of 28 benign prostatic hyperplasias. As outlined for chromosome Y losses, this observation does not exclude a role of chromosome 7 gains for tumor progression, it shows, however, that polysomy 7 cannot be used as a parameter of malignancy in prostate biopsies. Chromosome 7q deletions occur in about 20% of prostatic carcinomas [34, 40]. The most common site of deletion was mapped to 7q31.1 [34]. The finding of a higher prevalence of 7q31 deletions in node positive (4 of 7 deleted) than in node negative tumors (8 of 47 deleted; $p = 0.017$) would be consistent with a role of a tumor suppressor gene at 7q31.1 for prostate cancer progression (34).

2.3 Chromosome 8 alterations

Chromosome 8p deletions occur in 60–80% of primary carcinomas of the prostate [10, 25, 33, 35] and in up to 90% of prostatic intraepithelial neoplasms [14]. These suggests that a tumor suppressor gene on 8p is relevant for prostate cancer initiation and that the inactivation of this gene occurs in early stage tumors. Mapping studies have located at least two different putative tumor suppressor gene on 8p. One candidate gene has been mapped to 8p22–p21.3 [33], the other is more proximal at 8p21–p11.22 [33]. The same two loci have been identified in positional cloning studies on colorectal tumors [15], suggesting that tumor suppressor genes with a significant role for several tumors may reside on these loci.

A few extra copies of chromosome 8q (8q gain) is another frequent finding in prostate cancer. CGH studies have shown that these 8q overrepresentations are frequently large, often involving entire 8q [10]. The combination of 8p deletions and 8q gains in many tumors may reflect presence of isochromosomes 8q. While it is tempting to speculate that 8q gains could lead to overexpression of an oncogene on 8q, it has never been shown that a few extra copies of a chromosome or a part of a chromosome can lead to a detectable oncogene overexpression. In a previous FISH study we found that c-myc (located at 8q24) protein expression was even lower in bladder carcinomas with an 8q24 gain that in tumors without 8q24 overrepresentation [29]. This indicates that low level gains of chromosomal material does not necessarily lead to a significant overexpression of all genes on a given chromosome arm.

2.4 10q deletions

10q deletions have been described to occur in a number of different malignancies including gliomas, malignant melanoma and urinary bladder cancer [26]. In prostate cancer 10q deletions have been found in as many as 20% of tumors characterized by cytogenetics [26]. The observation of 10q deletions as the sole detectable aberration in 3 carcinomas of the prostate lead to the conclusion that 10q deletions may belong to the early alterations in prostate cancer development [4]. Mapping studies have suggested a tumor suppressor gene located between 10q23 and 10q25. Interestingly, this region contains MXI 1. MXI 1 negatively regulates Myc oncoprotein activity and thus potentially serves a tumor suppressor function. The finding of MXI 1 mutations in several prostatic carcinomas with 10q deletions raises the possibility that MXI 1 is a tumor suppressor gene with relevance for prostate cancer [12].

2.5 16q deletions

Deletions of 16q occur in 30 to 50% of clinically localized prostate carcinomas [9]. The potential tumor suppressor gene on 16q is unknown. A recent study on breast carcinomas suggests that two different tumor suppressor genes may reside on 16q, one on 16q22.1 and another on 16q24.3 [11]. The E-cadherin gene is a strong candidate to represent the tumor suppressor gene at 16q22.1. The E-cadherin gene contains codes for a cell surface glycoprotein adhesion molecule important in cell-cell interactions. It is postulated that reduced E-cadherin function facilitates tumor invasiveness due to a decreased cellular cohesiveness. However, a recent study by Cher et al. [9] suggest that the E-cadherin gene does not fall within the region of common deletion in prostate cancer. Other potential tumor suppressor genes include an adhesion regulator gene, located in the peritelomeric region of chromosome 16 [28]. This gene codes for a protein involved in cell-matrix adhesion. A loss of this protein might also lead to an increased tumor invasiveness.

2.6 Gene amplification

Although cytogenetic evidence for gene amplification (double minutes) has been found in short term cultures of prostate cancer cells [24], gene amplification is apparently rare in prostate cancer. Southern blot examinations had revealed no amplifications of cyclin D1, c-myc, and erbB-2 [23]. Also two recent CGH studies have – in contrast to the findings in many other solid tumors [18, 20] – not found circumscribed high level amplifications in primary prostate cancer [10, 39]. The androgen receptor which is found amplified in almost 30% of hormone resistant carcinomas [38] is the only gene for which amplification seems to occur at a relevant frequency in prostate cancer.

2.7 Other cytogenetic changes in prostate cancer

A number of other cytogenetic alterations have been repeatedly found in prostate cancer. CGH studies have described frequent losses at 2q, 5q, 6q, 9p, 13q, 17p, and 18q [8, 10, 39]. Gains of chromosomal material were most frequently found at 1q, 2p, 4q, and 17q [18, 10, 39]. Little is currently known on the genes involved in most of these areas. However, given the rapid improvements of molecular technologies designed to decipher the genetic background of cancer, it is likely that the target genes at these and other loci will be discovered soon.

References

(1) Alcaraz, A., S. Takahashi, J. A. Brown et al.: Aneuploidy and aneusomy of chromosome 7 detected by fluorescence in situ hybridization are markers of poor prognosis in prostate cancer. Cancer Res. *54* (1994) 3998–4002.

[2] Alers, J. C., P. J. Krijtenburg, K. J. Vissers et al.: Interphase cytogenetics of prostatic adenocarcinoma and precursor lesions: analysis of 25 radical prostatectomies and 17 adjacent prostatic intraepithelial neoplasias. Genes Chrom. Cancer *12* (1995) 241–250.

[3] Aly, M. S., P. D. Cin, d. V. W. Van et al.: Chromosome abnormalities in benign prostatic hyperplasia. Genes Chrom. Cancer *9* (1994) 227–233.

[4] Arps, S., A. Rodewald, B. Schmalenberger et al.: Cytogenetic survey of 32 cancers of the prostate. Cancer Genet Cytogenet, *66* (1993) 93–99.

[5] Bandyk, M. G., L. Zhao, P. Troncoso et al.: Trisomy 7: a potential cytogenetic marker of human prostate cancer progression. Genes Chrom. Cancer *9* (1994) 19–27.

[6] Baretton, G. B., C. Valina, T. Vogt et al.: Interphase cytogenetic analysis of prostatic carcinomas by use of nonisotopic in situ hybridization. Cancer Res *54* (1994) 4472–4480.

[7] Casalone, R., C. P. Granata, E. Minelli et al.: Significance of the clonal and sporadic chromosome abnormalities in non-neoplastic renal tissue. Hum. Genet. *90* (1992) 71–78.

[8] Cher, M. L., G. S. Bova, D. H. Moore et al.: Genetic alterations in untreated metastases and androgen-independent prostate cancer detected by comparative genomic hybridization and allelotyping. Cancer Res. *56* (1996) 3091–3102. issn: 0008–5472.

[9] Cher, M. L., T. Ito, N. Weidner et al.: Mapping of regions of physical deletion on chromosome 16 q in prostate cancer cells by fluorescence in situ hybridization (FISH). J. Urol. *153* (249–254.

[10] Cher, M. L., D. MacGrogan, R. Bookstein et al.: Comparative genomic hybridization, allelic imbalance, and fluorescence in situ hybridization on chromosome 8 in prostate cancer. Genes Chrom. Cancer *11* (1994) 153–162.

[11] Cleton-Jansen, A., E. Moerland, N. Kuipers-Dijkshoorn et al.: At least two different regions are involved in allelic imbalance in chromosome arm 16 q in breast cancer. Genes Chrom. Cancer *9* (1994) 101.

[12] Eagle, L. R., X. Yin, A. R. Brothman, et al.: Mutation of the MXI 1 gene in prostate cancer. Nat. Genet. *9* (1995) 249–255.

[13] Elfving, P., J. C. Cigudosa, R. Lundgren et al.: Trisomy 7, trisomy 10, and loss of the Y chromosome in short-term cultures of normal kidney tissue. Cytogenet. Cell Genet. *53* (1990) 123–125.

[14] Emmert, B. M., C. D. Vocke, R. O. Pozzatti et al.: Allelic loss on chromosome 8 p12–21 in microdissected prostatic intraepithelial neoplasia. Cancer Res. *55* (1995) 2959–2962.

[15] Fujiwara, Y., M. Emi, H. Ohata et al.: Evidence for the presence of two tumor suppressor genes on chromosome 8 p for colorectal carcinoma. Cancer Res. *53* (1993) 1172–1174.

[16] Hopman, A., F. Ramaekers, A. Raap et al.: In situ hybridization as a tool to study numerical chromosome aberrations in solid bladder tumors. Histochemistry *89* (1988) 307–316.

[17] Johansson, B., S. Heim, N. Mandahl et al.: Trisomy 7 in nonneoplastic cells. Genes Chrom. Cancer *6* (1993) 199–205.

[18] Kallioniemi, A., O. Kallioniemi, G. Citro et al.: Identification of novel DNA amplifications and deletions in primary bladder cancer by comparative genomic hybridization. Genes Chrom. Cancer *12* (1995) 213–219.

[19] Kallioniemi, A., O. Kallioniemi, D. Sudar et al.: Comparative genomic hybridization for molecular cytogenetic analysis of solid tumors. Science *258* (1992) 818–821.

[20] Kallioniemi, A., O. P. Kallioniemi, J. Piper et al.: Detection and mapping of amplified DNA sequences in breast cancer by comparative genomic hybridization. Proc. Nat'l Acad. Sci. USA *91* (1994) 2156–2160.

[21] Kallioniemi, O., A. Kallioniemi, W. Kurisu et al.: C-erbB-2 oncogene amplification in breast cancer analyzed by fluorescence in situ hybridization. Proc. Nat'l Acad. Sci. USA *89* (1992) 5321–5325.

[22] Konig, J. J., W. Teubel, D. J. Van de Ven et al.: Loss and gain of chromosomes 1, 18, and Y in prostate cancer. Prostate *25* (1994) 281–291.

[23] Latil, A., J. C. Baron, O. Cussenot et al.: Oncogene amplifications in early-stage human prostate carcinomas. Int. J. Cancer *59* (1994) 637–638.
[24] Limon, J., R. Lundgren, P. Elfving et al.: Double minutes in two primary adenocarcinomas of the prostate. Cancer Genet. Cytogenet *39* (1989) 191–194.
[25] Macoska, J. A., T. M. Trybus, W. A. Sakr et al.: Fluorescence in situ hybridization analysis of 8p allelic loss and chromosome 8 instability in human prostate cancer. Cancer Res. *54* (1994) 3824–3830.
[26] Mitelman, F.: Catalog of chromosome aberrations in cancer. Wiley-Liss, New York 1994.
[27] Powell, I., M. Tyrkus, E. Kleer: Apparent correlation of sex chromosome loss and disease course in urothelial cancer. Cancer Genet. Cytogenet *50* (1990) 97–101.
[28] Pullman, W., W. Bodmer: Cloning and characterization of a gene that regulates cell adhesion. Nature *356* (1992) 529.
[29] Sauter, G., P. Carroll, H. Moch et al.: Increased c-myc gene copy number in advanced bladder cancer. Am. J. Pathol. *146* (1995) 1131–1139.
[30] Sauter, G., T. Gasser, H. Moch et al.: DNA aberrations in urinary bladder cancer detected by flow cytometry and FISH. Urol. Res. *25* (1997) S 37–44.
[31] Sauter, G., H. Moch, U. Wagner et al.: Y Chromosome loss detected by FISH in bladder cancer. Cancer Gen. Cytogen. *82* (1995) 163–169.
[32] Schröck, E., S. Du Manoir, T. Veldman et al.: Multicolor spectral karyotyping of human chromosomes. Science *273* (1996) 494–497.
[33] Suzuki, H., M. Emi, A. Komiya et al.: Localization of a tumor suppressor gene associated with progression of human prostate cancer within a 1.2 Mb region of 8p22–p21.3. Genes Chromosomes Cancer *13* (1995) 168–174.
[34] Takahashi, S., A. L. Shan, S. R. Ritland et al.: Frequent loss of heterozygosity at 7q31.1 in primary prostate cancer is associated with tumor aggressiveness and progression. Cancer Res. *55* (1995) 4114–4119.
[35] Trapman, J., H. F. Sleddens, d. W. M. Van de Ven et al.: Loss of heterozygosity of chromosome 8 microsatellite loci implicates a candidate tumor suppressor gene between the loci D8 S87 and D8 S133 in human prostate cancer. Cancer Res. *54* (1994) 6061–6064.
[36] United Kingdom Cancer Cytogenetics Group: Loss of the Y chromosome from normal and neoplastic bone marrows. Genes Chrom. Cancer *5* (1992) 83–88.
[37] Van Dekken, H., J. Alers: Loss of chromosome Y in prostatic cancer, but not in stromal tissue. Cancer Genet. Cytogenet. *66* (1993) 131–132.
[38] Visakorpi, T., E. Hyytinen, P. Koivisto et al.: In vivo amplification of the androgen receptor gene and progression of human prostate cancer. Nat. Genet. *9* (1995) 401–406.
[39] Visakorpi, T., A. H. Kallioniemi, A. C. Syvanen et al.: Genetic changes in primary and recurrent prostate cancer by comparative genomic hybridization. Cancer Res. *55* (1995) 342–347.
[40] Zenklusen, J. C., J. C. Thompson, P. Troncoso et al.: Loss of heterozygosity in human primary prostate carcinomas: a possible tumor suppressor gene at 7q31.1. Cancer Res. *54* (1994) 6370–6373.

The regulation of prostatic growth

K. Griffiths

1. Introduction

Symptomatic clinical benign prostatic hyperplasia (BPH), resulting from uncontrolled growth of various cellular elements of the prostate, generally presents beyond the age of 50 years. The age factor seems to be implicated in the etiology of the disease and the majority of men greater than the age of 60 years will manifest some degree of enlargement of the prostate gland. A source of circulating testosterone would appear to be necessary for the pathogenesis of BPH as for the development of carcinoma of the prostate, but paradoxically both conditions occur when testicular function is declining [30]. Early castration, however, would seem to prevent the development of clinical BPH and of carcinoma of the prostate. Furthermore, the enlarged prostate regresses after orchiectomy or administration of LH-RH-analogues [24], again indicating that the abnormal growth is probably mediated by androgenic hormones.

With more recent reports [8] that a man at the age of 40 now has a 30% chance in his lifetime of requiring a prostatectomy for the treatment of symptomatic BPH compared to a value of 10% in 1968 and that 20 to 30% of men beyond the age of 50 have a latent carcinoma of the prostate, greater effort should be directed to a better understanding of the natural history of both diseases and to the molecular endocrinology within the cell, that is implicated in the growth regulatory processes of the prostate. Such a need becomes even more apparent in view of the knowledge that world-wide, life-expectancy continues to be extended. New approaches of therapy can only be derived from a greater understanding of the endocrine, biochemical and molecular processes associated with both conditions.

2. Essential concepts of the biological processes responsible for prostatic growth

2.1 Androgens and the prostate

The prostate gland is androgen dependent requiring a source of testosterone for its growth, development, differentiation and function. Removal of the source of testosterone by bilateral orchidectomy causes prostate involution. Subsequent administration of exogenous testosterone results in re-growth [13], although the gland only attains its original size. Essentially, the response of the gland to exogenous androgen does not cause prostatic overgrowth.

Within the prostate gland, testosterone is converted to 5-alpha-dihydrotestosterone (DHT) by the 5-alpha-reductase (5aR) enzyme system localised on the nuclear membrane. DHT preferentially binds to the androgen receptor (AR) and it is the DHT-AR-complex which associates with nucelotide recognition sequences on the genome, referred to as hormone response elements. These are adjacent to androgen responsive genes related as well to growth regulatory processes as to expression of secretory proteins such as PSA as to other cellular functions. In the prostate gland, therefore, DHT-AR directly modulates gene expression. The binding of DHT-AR with the specific enhancer sequences in the promoter region of the gene allows the complex to act as a ligand-dependent transcription factor and the DHT to modulate the activity of the androgen responsive genes (Fig. 1).

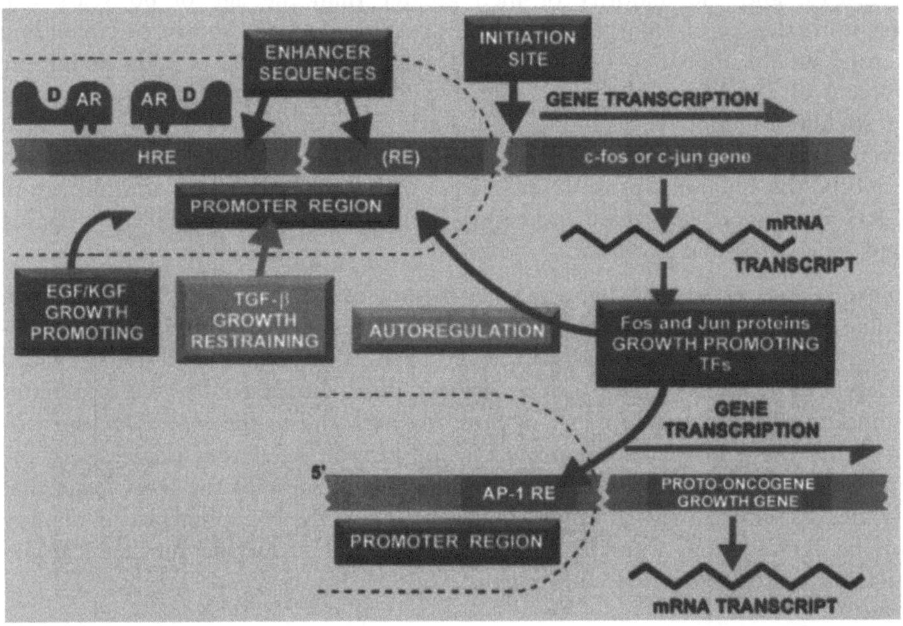

Fig. 1 A simplified representation of gene regulation, essentially, the androgen modulation of the expression of the c-fos or c-jun proto-oncogenes. The products encoded by the genes, the Fos and Jun products can act as growth promoting agents (TF) and as Fos/Jun dimer complexes, bind to another recognition site (AP-1 site) on other promoter regions of other genes concerned with growth. The Fos and Jun proteins are seen to auto-repress the expression of their own genes, which are also positively regulated by epidermal growth factor (EGF) and keratinocyte growth factor (KGF), and restrained by transforming growth factor-β (TGF-β).

2.2 Growth factors and the prostate

The androgens play an important role in regulating gene activity and the prostate growth. Isolated prostatic epithelial cells in culture do not, however, respond directly to androgens. But increased growth and proliferation is seen in response to growth factors such as epidermal growth factor (EGF), keratinocyte growth factor (KGF) and the insulin-like growth factors-I and -II (IGF-I and -II). These growth stimulatory factors appear to be mitogenic and directly stimulate cell proliferation, whereas DHT orchestrates these effects.

2.3 Stromal-epithelial inter-relationships

The prostate gland consists of a complex ductal system, with the ducts, lined by epithelial cells, embedded in a matrix of stromal components. The stromal elements are now recognised to play an important role in regulating various aspects of cellular function, including growth. As part of the normal growth regulatory processes of the prostate gland, androgen-mediated effects on the stromal compartment produce growth stimulatory factors which induce various signal transduction pathways that influence epithelial growth and differentiation. The AR is present in both the epithelial and stromal compartments, whereas the oestrogen receptor (ER) is predominantly located in the stromal elements [10]. Recent reports suggest that the effective 5aR enzyme system may be sited primarily in the stroma and that the epithelial 5aR is in the basal cells [27].

During the period of rapid growth of the prostate, between adolescence and the early 20s, when the human prostate attains its maximal size, the stromally-derived growth factors exercise a major role in epithelial cell proliferation and the laying down of the ducts in the stroma. Once the adult size is reached, homeostasis is established with growth inhibitory factors such as transforming growth factor-β (TGF-β), known to restrain epithelial cell proliferation. McNeal proposed a hypothesis that the pathogenesis of BPH related to the 're-awakening of the embryonic inductive capacity of the adult prostatic stroma' [19]. Either an over-active growth promoting effect, or an ineffective ']olecular shut down' restraining influence of TGF-β, may disturb the balance in the gland and lead to early microscopic BPH. Recently particular hormones and growth factors could be identified that appear to exercise important physiological roles in the prostate.

2.4 The role of KGF

Apart from EGF as a probable growth stimulatory factor, attention has more recently been directed to members of the fibroblast growth factor (FGF) family. In contrast to earlier studies the accumulated evidence now indicates that FGF-1 and FGF-2 stimulate the proliferation of stromal cells [16]. Studies of Chung Lee and co-workers suggest that DHT may enhance these effects of the FGF, possibly synergistically with oestradiol-17β promoting stromal cell hyperplasia [9]. In transgenic mice, the overexpression of the int-2 gene that encodes for another member of the FGF family, FGF-3, caused prostatic epithelial hyperplasia [21]. KGF, also a member of the FGF family (FGF-7), is mitogenic in cultures of normal prostatic epithelial cells, but does not promote stromal proliferation [32, 33]. The growth stimulatory effects of KGF are inhibited by TGF-β.

2.5 Regional variation of epithelial function in the ductal system

The inter-relationship between the epithelial and stromal compartments of the prostate is rather complex. Several studies have identified a distinct regional variation in the functional activity of epithelial cells in different regions of the rat prostate ductal system, described as proximal, intermediate or distal segments relative to the distance from the urethra [14]. Epithelial cells in the distal region are actively proliferating; the differentiated cells in the intermediate section, the major component, appear to be secretory, and the proximal segment is associated with programmed cell death, or apoptosis. This regional variation in morphology and function in the presence of normal circulating levels of testosterone suggests a corresponding regional variability in response to DHT and to growth regulatory factors such as KGF or TGF-β. The latter will be exercising a prominent role in the proximal segment, being recognised as an important growth regulatory factor in apoptosis [12], a process over which, DHT elicits an antagonistic influence. The data also implies that there may be a migration of epithelial cells from the distal proliferating region, along the duct, to the proximal segment where apoptosis prevails. The marked proliferative activity of the distal region may also relate to the situation in the human prostate, where prostatic intraepithelial neoplasia (PIN) is primarily recognised in the peripheral zone, the site of origin of approximately 70% of cancers.

2.6 Microscopic BPH

The concept of cellular homeostasis with regional variability in response to androgen, would suggest that the early phases in the development of symptomatic BPH, epithelial hyperplasia leading to microscopic BPH and nodular hyperplasia could be a consequence of the imbalance in the equilibrium between the biological effects of the growth regulatoriy factors, dysfunctional cell migration or ineffective processes related to apoptosis. The early phases in the development of symptomatic BPH would certainly appear to be related to androgen promotion or modulation. These early phases are associated primarily with epithelial cell proliferation.The major proportion of the nodules identified at an early stage are indeed epithelial, and are located in the transition zone. Possibly, with increasing genetic instability, this epithelial cell proliferation can be associated with the development of cancer in the transition zone.

3. Developing concepts: new approaches

3.1 The epithelial-stromal inter-relationship

Androgens are clearly the principal determinants of prostate growth, but their biological effects and tissue responses are effected through various intermediary intrinsic factors. The inter-relationship between stromal and epidermal elements would also appear to be pivotal to the growth regulatory processes of the prostate. The concept of a regional heterogeneity in the ductal system of the rat prostate emphasises the complexicity of these processes and the need to identify these intermediary factors, the key messengers involved in the maintenance of cellular homeostasis and ultimately, in disease processes. All three glandular segments receive the same level of androgenic support, yet differing patterns of cellular behavior are observed (Fig. 2).

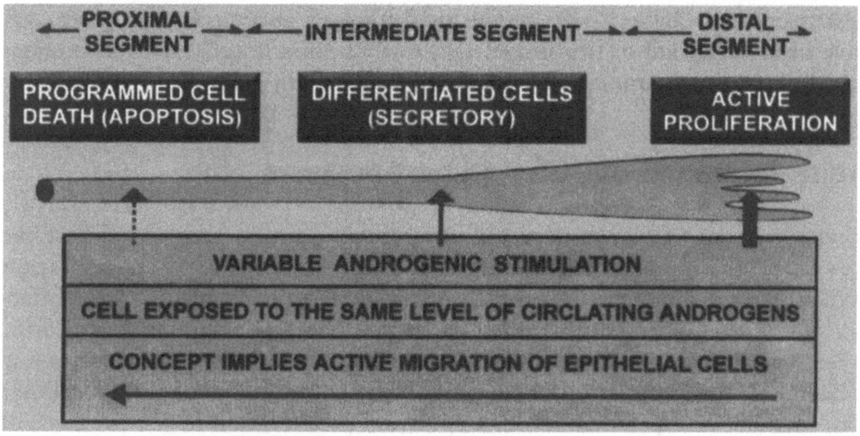

Fig. 2 Variable epithelial cell function along the ducts.

This is imcompatible with the belief that DHT is the sole mediator of prostatic growth control and function.

Certain questions have to be posed:

- Is the variable response to androgen due to a corresponding variable composition of the stromal cells underlying the epithelial cells in the different segments?
- Is there a regional variability in the production of growth regulatory factors in response to androgenic stimulation?
- Locally regulated cell-cell interactions with regional variability in stromal-epithelial inter-relationships should now be considered. The question as to the nature of these differing cellular responses relating to the production of, and response to the various growth regulatory factors requires elucidation.

3.2 The stromal cells

There is evidence that the stroma surrounding the ducts does indeed manifest regional heterogeneity with regard to the distribution of smooth muscle-like cells. In the proximal segment of the duct, there are multiple layers of smooth muscle cells, whereas in the distal region, a single layer of cells is observed. Prostatic fibroblasts are reported to be evenly distributed throughout the stromal elements of the duct. The androgen-mediated production of stimulatory factors such as KGF by the fibroblasts would promote epithelial cell proliferation, whereas other factors from the smooth muscle cells would restrain proliferation. TGF-β, considered to be directly associated with the processes of apoptosis, has generally been found to be located in the stromal elements and particularly in the actin-positive smooth muscle cells [15]. It must be stated that the factors which control the growth and differentiation of the stromal cell population are not well understood. It is currently accepted that FGF-2 is the principal component promoting fibroblast growth in culture [26]. DHT itself has an effect on the proliferation of prostatic fibroblasts in culture only in association with FGF-2. Oestrogens may have a role in the inhibition of apoptosis. Moreover, AR and ER have been identified in fibroblasts and smooth muscle cells, with androgens inhibiting and oestrogens stimulating smooth muscle growth [29].

3.3 A major influence of TGF-β in homoestasis

Apoptosis occurs in an androgen-deprived environment and is seen as part of the normal growth regulatory balance within the prostate gland [17]. Therefore, apart from growth stimulatory factors, it is equally reasonable to consider that an ineffective control of the programmed cell death, such that there is a failure to kill cells, could also lead to prostate enlargement. Oestrogens may be concerned in restraining apoptosis. Some of the interactions being involved in the maintenance of cellular homeostasis, include TGF-β which inhibits epithelial proliferation in the presence of a growth stimulatory factor such as KGF or FGF, possibly the environment of the secretory intermediate segment of the duct. In the proximal segment, TGF-β would promote cell death in the absence of KGF and without the antagonistic influence of androgen. In the distal region, KGF in the absence of TGF-β, would promote epithelial proliferation.

Again, questions are posed:

- Does TGF-β influence the production of KGF by stromal fibroblasts in the presence of adequate levels of the modulating androgens, in order to regulate epithelial proliferation?
- From which compartment is the TGF-β produced and does it regulate the expression of androgen-responsive proto-oncogenes implicated in encoding for KGF?
- Does DHT regulate the expression of the TGF-β gene?

3.4 Growth regulatory imbalance

The mitogenic activity of KGF is characterised by an effect on cultures of normal prostatic epithelial cells and its lack of effect on stromal cells. In contrast, FGF-2 is mitogenic for prostatic stromal cells [26], but has no effect on epithelial cells. Both cell types, however, synthesise FGF-2 and an elevated expression of FGF-2 in association with BPH is reported. Particularly interesting is the marked complexity of the mechanisms by which some of the FGF receptor (FGF-R) genes are regulated and expressed. Four FGF-R genes have been recognised.

The FGF-R1 (flg) gene encodes a receptor protein that is normally recognised in stromal cells [32, 33]. The prostate epithelial cells express a splice variant of the FGF-

R2 (bek) gene, a receptor protein referred to as FGF-R2-exon IIIb, which is a receptor specifically recognising KGF. The splicing enzymes structure the primary mRNA to produce the mature mRNA transcript. The splicing variants are produced by variation in the structure of the exons which are built into the mature mRNA, with the consequent synthesis of receptors with structural heterogeneity. Such differing receptor phenotypes can elicit a stimulatory, or alternatively an inhibitory signal, in response to different FGFs.

In the prostate, the KGF-specific FGF-R2-exon IIIb receptor is expressed in the slowly growing, non-metastatic and well differentiated Dunning R-3327 PAP prostatic tumours. This receptor is not present in the rapidly growing, undifferentiated cancer, although the FGF-R2-exon IIIc with its high affinity for FGF-2 was recognised in this tumor. The FGF-R1 gene which encodes the receptor nomally recognised in the stromal cells, was also identified in the rapidly growing cancer [32, 33]. Such alterations in the specificity of an FGF-R that makes it unable to recognise KGF, but avidly bind the FGF-2, may underlie the early phases in the imbalance of growth regulatory processes within the prostate that ultimately lead to epithelial hyperplasia, and eventually to neoplasia. Accumulating evidence indicates that KGF has an important role as a paracrine mediator involved in the stromal-epithelial inter-relationship and as a regulator of prostatic growth.

3.5 Transition zone cancer

It was originally believed that the major proportion of prostatic cancers originated in the peripheral zone of the gland. The transition zone was accepted as the site of origin of BPH and the general consensus has been that prostatic cancer and BPH represented independent disorders with distinct aetiologies.Recent reports indicate that although 70% of cancer do originate in the peripheral zone, approximately 25% develop in the centrally located transition zone and are identified as 'incidental carcinoma' in transurethrally resected tissue at prostatectomy (TURP) for bladder outlet obstruction [3]. Transition zone cancer and BPH show an increasing incidence with age, have related natural history and both are androgen associated conditions that respond to androgen withdrawal therapy. Cancer is found incidentally in approximately 10% of TURP specimen, with an incidence which is age-dependent.

Resected tissue also contains lesions resembling low grade acinar carcinoma, referred to as atypical adenomatous hyperplasia (AAH), which could be considered analogous to PIN considered a premalignant lesion in the peripheral zone [2]. Both AAH and PIN represent early stages in the escape of epithelial cell proliferation from growth regulation.

A critical re-evaluation of the epidemiology, pathology and moelcular biology of prostatic disease suggests a potential relationship between microsopic BPH and transition zone cancer [3]. The same growth regulatory factors seem to be implicated in the multi-step progression of both diseases. It would be supposed that the early phases of BPH, essentially microscopic lesions of epithelial hyperplasia leading to nodular hyperplasia, are cellular conditions that could result in genetic mutations consistent with preneoplasia.

3.6 Nuclear matrix

Information is accumulating to support this concept of progressive changes. Cells which respond to external stimuli such as these induced by KGF or FGF depend, not only on the presence of cell membrane-located receptor proteins, but also on the particular activity of the signal tranduction pathways, as well as the intracellular structural matrix of the target cell. The nuclear matrix is a cell-type specific framework implicated in the organisation of DNA and in the regulation of gene transcription. The components of the nuclear matrix have been shown to be involved in steroid hormone regulation of gene expression, with specific association between steroid receptors and matrix components [1]. Other factors influencing transcriptional processes such as the c-myc proto-oncogene product interact with the matrix [7]. Clearly mutation-induced alterations in the composition of the matrix could influence cell behaviour, which would be reflected in changes in the protein constituents of the matrix. Recent studies, involving the 2-dimensional electrophoretic analysis of the matrix of normal, BPH and prostatic cancer specimen support the concept of an additive multi-step progression from the normal, through BPH to carcinoma, as opposed to distinct independent lines of progression for each disorder [23]. The immunocytochemical analysis of the keratins of prostatic tissue also supports this concept.

3.7 Cross-talk between signalling pathways

Fundamental to the hormone responsiveness of the prostate is the inter-relationship between signalling pathways, those effecting a biological response through steroid receptor proteins and those activated by the binding of the growth regulatory factors to cell membrane receptors. The interaction between DHT-AR with the AREs allows the complex to modulate the transcription of particular genes depending on the function of the cell, i. e. the complexes may suppress as well as activate gene expression. DHT-AR may modify the spatial orientation of the DNA strands to facilitate the easier access to the enhancer sequences of the promoter region of the genes of other transcription factors. Such factors as the Fos and Jun proteins, encoded by the c-fos and c-jun proto-oncogenes, are immediately concerned in growth regulation and their intracellular production is enhanced by growth factors such as KGF and FGF. The Fos and Jun proteins bind as heterodimers to the AP-1 recognition sites (Fig. 1) and probably can interact with the AR proteins. The cross-talk between the steroid receptor controlled pathways and those activated by growth factors, obviously a finely-tuned system, is the very fulcrum of growth regulatory processes.

3.8 Neuroendocrine cells

The neuroendocrine cells of the prostate, which are irregularly distributed throughout the acinar tissue, contain and probably can secrete serotonin, calcitonin, bombesin-like proteins and other types of proteins such as somatostatin, human chorionic gonadotrophin and thyroid stimulating hormone. These cells could exercise a paracrine, local regulatory role, possibly relating to growth, but also to the control of secretory processes of the gland. Focal neuroendocrine differentiation in carcinoma of the prostate is virtually ubiquitous and relates to poor prognosis [6]. The role of these cells in carcinogenesis of the prostate also requires investigation. Serotonin has been reported to be mitogenic in certain tissues and appears to exercise a permissive role in relation to androgen action, regulating prostatic epithelial cell proliferation [25]. Synergistically with other growth regulatory factors, serotonin stimulates fibroblast and smooth muscle mitogenesis.

3.9 Oestrogens and the human prostate gland

It has been known for a long time that oestrogens are required for the growth of the male accessory tissue. The precise role of these hormones still remains to be identified. Nevertheless, it is generally accepted that oestrogens modify androgen metabolism of the prostate. Oestradiol, administered with androgens, causes prostatic enlargement, essentially glandular hyperplasia, in dogs [31]. In canine BPH, the nuclear AR levels are specifically increased by oestrogen [20]. The human prostate gland accumulates oestrogen [11]. The mean levels of oestrone and oestradiol are significantly higher in BPH stromal tissue relative to BPH epithelial tissue, as well as to epithelial and stromal tissue in normal prostate.

Some more recent studies also support a role for oestrogens: So, for example, the synergistic effect of administered oestradiol and testosterone is reported to have promoted the growth of the rat prostate gland more than when testosterone was administered separately [28]. Oestrogens also seem to increase the responsiveness to androgen, of human stromal cells from BPH by regulation of AR levels [5]. Oestradiol, but not DHT, acting in association with SHBG, produces an 8-fold increase in the intracellular levels in human BPH tissues [22], cAMP being concerned in signal transduction pathways. DHT, which restrains the binding of oestradiol, inhibits this effect and this regulatory process is stated to be localised primarily in the stroma.

3.10 Diet and prostate disease

The marked differences in the incidence of both prostatic and breast cancer between the East and the West (Fig. 3) directs attention to factors in the West, such as fat products, which may be causal, or alternatively, to constituents of the diet of the people of the East that may restrain the processes of carcinogenesis. Furthermore, the incidence of obstructive BPH in the East follows a similar pattern to that observed for prostatic cancer. The studies of migrating peoples from China and Japan, through Hawai to mainland USA, indicate that the migrants, in a few generations, show a mortality rate for prostatic cancer that is close to that of the indigenous American people, suggesting that dietary rather than genetic factors are implicated. It is interesting to note that the incidence of microscopic BPH and of latent carcinoma of the

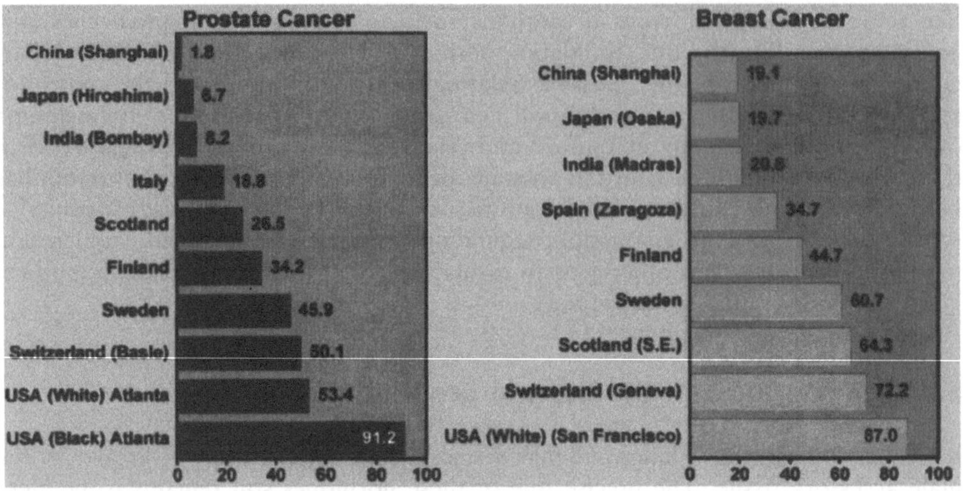

Fig. 3 Simple illustration of the differences in the incidence rates from cancers of the prostate and the breast between countries in the east and the west.

prostate is the same for all races [4]. It is suggested, therefore that constituents of the Asian diet and to some extent, of the diet in the Mediterranean Region, restrain the later phases in the pathogenesis of obstructive BPH, as well as the progression of the latent carcinoma to the malignant phenotype.

Certain products of vegetables, fruit, whole grain and soya are metabolised by the normal gut microflora to particular compounds, the isoflavonoids and the lignans, in body fluids such as urine or prostatic fluid. These compounds, such as enterolactone and enterodiol and the isoflavonoids such as genistein and daidzein, are weak oestrogens and act biologically like the weak oestrogen tamoxifen, an antioestrogen. With regard to the prostate, these compounds have been reported to have the following weak characteristics:

- antioestrogenic,
- 5alpha-reductase inhibitors
- aromatase inhibitors,
- tyrosine specific protein kinase inhibitors, which would influence the effects of growth factors on cells
- angiogenesis inhibitors
- a possible promoting influence on the expression of the E-cadherin gene.

With oestrogens being important in the maintenance of prostatic homeostasis and in disease processes dietary weak oestrogens could therefore exercise a role in restraining the development of prostatic disease specifically, or western diseases generally.

3.11 The insulin-like growth factor axis

The complexity and sophistication of these steroid hormone modulated gene regulatory processes is well recognised. Concepts have been outlined regarding the type of imbalance that could lead to overactive cellular proliferation, hyperplasia and gland enlargement. The cross-talk between these steroid receptor-directed signalling pathways and those promoted by the regulatory growth factors must be intimately involved in these growth processes with the steroid receptor protein being essential and pivotal. Even at extremely low levels, the receptor protein exercises a major influence alongside the transcription factors such as the Fos and Jun proteins. Anomolous expression of such nuclear proto-oncogenes as c-myc, c-jun and c-fos could well be involved in the early phases that lead to the escape of cells from the delicate growth regulatory influence of steroid hormones, particularly the androgens, within the prostate gland.

The role of IGF-I and -II, however, has to be considered together with their family of high affinity binding proteins which regulate the availability of the IGFs to the cell membrane receptors (IGF-Rs). The proliferation of prostatic epithelial cells is stimulated by IGFs. These cells have been reported to express IGF-I receptor protein. IGF-II is produced by prostatic fibroblasts in culture and is considered to be implicated in the foetal development of the urogenital sinus [18]. Seminal plasma contains the IGFs and some of the IGF-BPs. The IGF-BP3 is also produced by fibroblasts under TGF-β control. The IGF-BP adjacent to the IGF receptor on the cell membrane avidly binds the IGF. The conformational change induced on the IGF-BP by a specific proteinase, possibly PSA, decreases the affinity of the IGF-BP such that the IGF dissociates and binds to the IGF-R, thereby inducing the positive growth stimulatory signals to promote epithelial cell proliferation. The exact role within the growth regulatory processes of the prostate remains to be elucidated.

Acknowledgements: The contents of this communication were the basis of a report to the 3rd International Consultation on BPH (held in Monte Carlo, 1995).

References

[1] Barrack. E. R., D. S. Coffey: The specific binding of oestrogens and androgens to the nuclear matrix of sex hormone responsive tissues. J. Biol. Chem. *255* (1980) 7265–7275.

[2] Bostwick, D. G., M. K. Brawer: Prostatic intra-epithelial and early invasion in prostatic cancer. Cancer *59* (1987) 788–794.

[3] Bostwick, D. G., F. S. Barcells, W. H. Cooner et al.: Benign prostatic hyperplasia (BPH) and cancer of the prostate. In.: A. T. K. Cockett, S. Khoury, Y. Aso et al. (eds.): The First International Consultation on Benign Prostatic Hyperplasia (BPH), pp 139–159. SCI Paris 1991

[4] Breslow, N. E., C. W. Cahn, G. Dhom et al.: Latent carcinoma of the prostate at autopsy in seven areas. Int. J. Cancer *20* (1977) 68–688.

[5] Collins, A. T., B. Zhiming, K. Gilmore et al.: Androgen and oestrogen responsiveness of stromal cells derived from the human hyperplastic prostate: oestrogen regulation of the androgen receptor. J. Endocrinol. *143* (1994) 269–277.

[6] Di Sant'Agnese, P. A.: Neuroendocrine differentiation in carcinoma of the prostate. Cancer *70* (1992) 254–268.

[7] Eisenman, R. N., C. Y. Tachibana, M. D. Abrams et al.: V-myc and c-myc encoded proteins are associated with the nuclear matrix. Mol. Cell Biol. *4* (1985) 114–126.

[8] Glynne, R. J., E. W. Campion, G. R. Bouchard et al.: The development of benign prostatic hyperplasia among volunteers in the normative aging study. Am. J. Epidemiol. *121* (1985) 78–90.
[9] Griffiths, K., H. Akaza, C. L. Eaton et al.: Hormones, growth factors and benign prostatic hyperplasia (BPH). In: A. T. K. Cocket, Y. Aso, C. Chatelain et al. (eds.): Proceedings of The International Consultation of Benign Prostatic Hyperplasia (BPH), pp 25–49. SCI Paris 1991.
[10] Krieg, M., W. Bartsch, M. Thomsen et al.: Androgens and estrogens: their interaction with stroma and epithelium of human benign prostatic hyperplasia and normal prostate. J. Steroid Biochem. *19* (1983) 155–161.
[11] Krieg, M., R. Nass, S. Tunn: Effect of aging on endogenous level of 5alpha-dihydrotestosterone, testosterone, estradiol, and estrone in epithelium and stroma of normal and hyperplastic human prostate. J. Clin. Endocrinol. Metab. *77* (1993) 375–381.
[12] Kyprianau, N., J. T. Isaacs: Expression of transforming growth factors in the rat ventral prostate during castration-induced programmed cell death. Mol. Endocrinol. *3* (1989) 1515–1522.
[13] Lee, C.: Physiology of castration-induced regression in rat prostate. In: G. P. Murphy, A. A. Sandbreg, J. P. Karr (eds.): The Prostatic Cell: Structure and Function, pp 145–159. Alan R. Liss Inc., New York 1981
[14] Lee, C., J. A. Sensibar, S. M. Dudek et al.: Prostatic ductal system in rats: Regional variation in morphological and functional activities. Biol. Reprod. *43* (1990) 1079–1086.
[15] Lee, C., J. M. Kozlowski, J. T. Grayhack: Etiology of benign prostatic hyperplasia. Urol. Clinics North Amer. *22* (1995) 237–246.
[16] Mansson, P. E., P. Adams, M. Kan et al.: Heparin binding growth factor gene expression and receptor characteristics in normal rat prostate and two transplantable rat prostate tumors. Cancer Res. *49* (1989) 2485–2494.
[17] Martikainen, P., N. Kyprianou, J. T. Isaacs: Effect of transforming growth factor-β on proliferation and cell death of rat prostatic cells. Endocrinology *127* (1990) 2963–2968.
[18] Martinoli, M. G., G. Pelletier: Dihydrotestosterone (DHT) regulation of insulin-like growth factor II mRNA in neonatal mice. Peptides *12* (1991) 1267–1271.
[19] McNeal, J. E.: Origin and evolution of benign prostatic enlargement. Invest. Urol. *15* (1978) 340–345.
[20] Moore, R. J., J. M. Gazak, J. D. Wilson: Regulation of cytoplasmic dihydrotestosterone binding in dog prostate by 17β-estradiol. J. Clin. Invest. *63* (1979) 351–357.
[21] Muller, W. J., F. S. lee, C. Dickson et al.: The int-2 gene product acts as an epithelial growth factor in transgenic mice. EMBO J. *9* (1990) 907–913.
[22] Nakhla, A. M., M. S. Khan, N. P. Romas et al.: Estradiol causes the rapid accumulation of cAMP in human prostate. Proc. Natl. Acad. Sci. USA *91* (1994) 5402- 5405.
[23] Partin, A. W., R. H. Getzenberg, M. J. Carmichael et al.: Nuclear matrix protein patterns in human benign prostatic hyperplasia and cancer. Cancer Res. *53* (1993) 744–746.
[24] Schroeder, F. H., M. Westerhof, R. J. L H. Bosch et al.: Benign prostatic hyperplasia treated by castration or the LH-RH analogue busereline : a report on six cases. Eur. Urol. *12* (1986) 318–321.
[25] Seuwen, K., J. Pouyssegur: Serotonin as a growth factor. Biochem. Pharmacol. *39* (1990) 985–990.
[26] Sherwood, E. R., C. J. Fong, C. Lee et al.: Basic fibroblast growth factor : A potential mediator of stromal growth in the human prostate. Endocrinology *130* (1992) 2955–2963.
[27] Silver, R. I., E. L. Wiley, D. L. Davis et al.: Expression and regulation of steroid 5 alpha reductase 2 in prostate disease. J. Urol. *152* (1994) 433–437.
[28] Suzuki, K., Y. Takezawa, T. Suzuki et al.: Synergistic effect of estrogen with androgen on the prostate – effects of estrogen on the prostate of androgen-administered rats and 5alpha-reductase activity. Prostate *25* (1994) 169–176.

[29] Toney, T. W., B. J. Danzo: Developmental changes in and hormonal regulation of estrogen and androgen receptors present in the rabbit epididymis. Biol. Reprod. *39* (1988) 818–828.

[30] Vermeulen, A.: Testicular hormone secretion and aging in males. In: J. T. Grayhack, J. D. Wilson, M. J. Scherbenske (eds): Benign prostatic hyperplasia. NIAMDD Workshop Proceedings, pp 177–182. NIH Publ., No. (NIH) 76–1113, Bethesda 1976.

[31] Walsh, P. C., J. D. Wilson: The induction of prostatic hypertrophy in the dog with androstanediol. J. Clin. Invest. *57* (1976) 1093–1097.

[32] Xu, J., M. Nakahara, J. W. Crabb et al.: Expression and immunocytochemical analysis of rat and human fibroblast growth factor receptor (flg) isoforms. J. Biol. Chem. *267* (1992) 17792–17803.

[33] Yan, G., F. Wang, Y. Fukabori et al.: Expression and transforming activity of a variant of the heparin-binding fibroblast growth factor receptor (flg) gene resulting from splicing of the alpha exon at an alternate 3'-acceptor site. Biochem. Biophys. Res. Commun. *183* (1992) 423–430.

Current concepts of oncogenes in prostate cancer

D. M. Peehl

Introduction

The search for oncogenes in prostate cancer began in the early 1980's, after it was recognized that certain oncogenes such as *ras* were commonly activated in some other types of cancers [44]. The initial results were disappointing because oncogenes which were frequently involved in other cancers, including *ras*, *myc*, and *sis*, did not appear to have a major role in prostate cancer. Recent findings, however, have led to a re-evaluation of the activity of *ras* and *myc* in prostate cancer. New technologies have also led to the identification of putative novel oncogenes in prostate cancer, and powerful new methods of genetic analysis promise to reveal other genes whose expression confers a malignant phenotype on a cell in a dominant fashion.

Ras

Most studies agree that *ras* mutations are quite infrequent in prostate cancers of American males [7, 13, 21, 26]. In contrast, *ras* mutations in prostate cancers of males in Japan seem to occur fairly often, perhaps at an overall rate of approximately 25% [1, 16, 47]. It is interesting to consider that this difference may reflect different etiologic agents causing prostate cancer or its progression in the two countries. The increased frequency of *ras* mutations in prostate cancers of the Japanese population may provide a basis for therapeutic strategies targeting *ras* oncogenes.

Although *ras* is rarely mutated in prostate cancers of American men, a recent immunohistochemical analysis of *ras* expression in normal and malignant prostatic tissues indicated that *ras* expression was elevated in cancer [5]. An earlier publication showed the same trend [46], but that study was discredited on the basis of nonspecificity of the antibody used [30]. The results of the latest study suggest that levels of *ras* expression in prostate cancer should be re-examined with specific antibodies that are now available. Although *ras* is not often mutated, it appears that increased signaling by *ras* could have a role in prostate cancer. Strategies to block *ras* signaling, therefore, could have therapeutic efficacy.

Ras has become a valuable tool in the development of experimental models of prostatic carcinogenesis. Normal human prostatic epithelial cells, transformed to immortality in culture with SV40, have been further transformed to tumorigenicity with *ras* oncogenes [24, 27]. Rat prostatic cancer cells with low metastatic potential became highly motile and metastatic when transfected with a *ras* oncogene [25]. Analysis of the phenotypes of these cells as an *in vitro* model of carcinogenesis may prove relevant to processes that occur during the initiation and progression of prostate cancer. In particular, these cells have been used to identify changes in response to growth-stimulatory or-inhibitory factors which correlate with transformation and malignant potential [27].

Ras has also been used in an *in vivo* mouse prostate reconstitution model in which cancer develops [42]. Embryonic urogenital sinus mesenchyme and epithelium are separated and infected with *ras* and *myc in vitro*, then are recombined and implanted under the kidney capsule. If *ras* alone is present, dysplasia occurs in the epithelium but no progression to cancer is seen. In another model in which SV40 expression is targeted to the prostate in transgenic mice, approximately 20% of the dysplastic lesions that developed were found to have mutated *ras* [38]. It appears that abnormal *ras* expression contributes to the development of prostatic dysplasia.

In the prostate reconstitution model, the expression of both *ras* and *myc* in the epithelium results in the appearance of adenocarcinomas [42]. Cancers which develop in the SV40 transgenic mice have *ras* mutations at a frequency of about 30% [38]. It is hoped that these models, which involve oncogenesis by *ras* but also additional genetic changes, will prove useful for screening new chemotherapeutic or chemopreventive agents for prostate cancer.

Myc

Our concept of the involvement of *myc* in prostate cancer has also evolved in recent years. Early studies hinted at the overexpression of *myc* in prostate cancer, but since the comparisons of normal and malignant tissues were done with whole tissues, it was difficult to evaluate the specific contribution of epithelial cells to the total levels of *myc* [6, 10]. Recent molecular studies demonstrate that *myc* expression is indeed elevated in prostate cancer, seemingly as a result of gene amplification. The biologic significance of this phenomenon is not known, but *myc* expression in the afore-described prostate reconstitution model causes hyperplasia [42]. As mentioned in the previous section, *ras* plus *myc* result in malignancy in this model. In an *in vitro* prostatic cell transformation model in which the introduction of SV40 DNA leads to immortality, subclones of the SV40-transformed cells were noted to contain amplified *myc* genes [8]. It will be interesting in this model system to see whether amplification of *myc* contributes to the development of a malignant phenotype.

C-erbB-2

C-erbB-2 is another oncogene whose expression has been rather intensely investigated in the prostate in recent years. The reported results have been somewhat confusing, possibly due to the use by different investigators of diverse immunohistochemical techniques, but a pattern may be emerging. Originally found by some investigators only in prostatic adenocarcinomas [17, 34, 51], *c-erbB-2* expression was subsequently noted by others in benign prostatic hyperplasia (BPH) and normal tissues [11, 12, 20, 23, 31]. Bostwick [3] summarized the findings to date as follows. *C-erbB-2* expression in normal glands is localized primarily to the basal epithelial cell layer. Basal cells of dysplastic epithelia retain expression of *c-erbB-2*, but luminal secretory cells in dysplastic lesions also express *c-erbB-2*. In adenocarcinomas, *c-erbB-2* is expressed by a majority of cells.

The pattern of expression of *c-erbB-2* suggests that *c-erbB-2* correlates with proliferative potential rather than aberrant regulation. The fact that the *c-erbB-2* gene is not amplified in prostate cancer further supports this concept [17, 51]. The prognostic significance of *c-erbB-2* expression in prostate cancer, if any, is still unclear [11, 17, 20, 32, 34].

Human papillomavirus (HPV)

HPV is a proven etiologic agent in genital warts and several cancers of the genitourinary tract [45]. This virus has also been proposed as a factor in the development of both benign prostatic hyperplasia (BPH) and prostatic adenocarcinoma. Detection rates ranging from 0 to 100% have been found in a large number of investigations evaluating the presence of HPV DNA in prostatic tissues using polymerase chain reaction (PCR) technology [9, 14, 19, 22, 33, 35, 36, 40, 43, 50]. A recent publication by Terris and Peehl indicates that these divergent results may depend on the primer set utilized for detection [41]. At any rate, no definitive role for HPV in prostate cancer has yet emerged.

Similar to other oncogenic agents, however, HPV has been a useful research tool and has been used to immortalize prostatic epithelial cells derived from normal or malignant tissues [4, 28, 29, 49]. Additional genetic changes have been introduced or induced and HPV-transformed cell lines have been converted to tumorigenicity [27, 28, 29, 48], providing an additional model of *in vitro* carcinogenesis. One tumorigenic cell line created by treating HPV-immortalized normal prostatic epithelial cells with a chemical carcinogen has a deletion in chromosome 8p [29], similar to a majority of human prostatic adenocarcinomas. Several other immortal cell lines created by transforming cells obtained from prostatic adenocarcinomas with HPV oncogenes also retained the 8p-deletion characteristic of the tumors of origin [4]. Therefore, characterization of the phenotype of these cells containing similar genetic changes to those of prostate cancers may reveal important information about the behavior of prostate cancer.

Novel oncogenes

Application of new techniques such as differential display [2, 18] or comparative genomic hybridization [15] has the potential to identify new oncogenes in prostate cancer which have previously gone unrecognized. One putative new prostatic oncogene discovered from differential display is PTI-1 [37]. This gene was shown to be expressed in 7 of 8 prostate cancers examined by Northern analysis, but not in several samples of benign tissues. Expression of PTI-1 is not prostate-specific, but may be prostate cancer-specific compared to normal prostatic tissues. The 3′-end of the PTI-1 gene is homologous to a truncated, mutated form of human elongation factor-1α, whereas the 5′-end is related to a ribosomal gene of mycoplasma. The unusual nature of this gene raised the concern that it was artifactual or perhaps present as a result of mycoplasma contamination, but further investigation by the original finders of this gene alleviate these concerns [39]. They have also demonstrated that PTI-1 transcripts can be detected in prostate carcinoma patient blood samples, suggesting a new basis for diagnosing prostate cancer or its progression.

Summary

The 1990s have seen a shift in direction from the search for oncogenes in prostate cancer to investigation of the role of tumor suppressor genes in the development of this malignancy. Nevertheless, both types of events – activation of oncogenes and loss of tumor suppressor genes – are believed to be essential for the development of cancers. Using novel and improved technologies, investigators are finding new roles for old oncogenes as well as identifying novel genes with potential oncogenic activity.

References

[1] Anwar, K., K. Nakakuki, T. Shiraishi et al.: Presence of ras oncogene mutations and human papillomavirus DNA in prostate carcinomas. Cancer Res *52* (1992) 5991–5996.

[2] Bauer, D., H. Muller, J. Reich et al.: Identification of differentially expressed mRNA species by an improved display technique (DDRT-PCR). Nucleic Acids Res. *21* (1993) 4272–4280.

[3] Bostwick, D. G.: c-erbB-2 oncogene expression in prostatic intraepithelial neoplasia: mounting evidence for a precursor role. J. Natl. Cancer Inst. *86* (1994) 1108–1109.

[4] Bright, R. K., C. D. Vocke, M. R. Emmert-Buck et al.: Generation and genetic characterization of immortal human prostate epithelial cell lines derived from primary cancer specimens. Cancer Res. *57* (1997) 995–1002.

[5] Bushman, E. C., R. N. Nayak, W. Bushman: Immunohistochemical staining of ras p21: staining in benign and malignant prostate tissue. J. Urol. *153* (1995) 233–237.

[6] Buttyan, R., I. S. Sawczuk, M. C. Benson et al.: Enhanced expression of the c-myc protooncogene in high-grade human prostate cancers. Prostate *11* (1987) 327–337.

[7] Carter, B. S., J. I. Epstein, W. B. Isaacs: Ras gene mutations in human prostate cancer. Cancer Research *50* (1990) 6830–6832.

[8] Degeorges, A., F. Hoffschir, O. Cussenot et al.: Recurrent cytogenetic alterations of prostate carcinoma and amplification of c-myc or epidermal growth factor receptor in subclones of immortalized PNT1 human prostate epithelial cell line. Int. J. Cancer *62* (1995) 724–731.

[9] Effert, P. J., R. A. Frye, A. Neubauer et al.: Human papillomavirus types 16 and 18 are not involved in human prostate carcinogenesis: analysis of archival human prostate cancer specimens by differential polymerase chain reaction. J. Urol. *147* (1992) 192–196.
[10] Fleming, W. H., A. Hamel, R. MacDonald et al.: Expression of the c-myc protooncogene in human prostatic carcinoma and benign prostatic hyperplasia. Cancer Res. *46* (1986) 1535–1538.
[11] Fox, S. B., R. A. Persad, N. Coleman et al.: Prognostic value of c-erbB-2 and epidermal growth factor receptor in stage A1 (T1a) prostatic adenocarcinoma. Br. J. Urol. *74* (1994) 214–220.
[12] Giri, D. K., R. Pal, S. N. Wadhwa et al.: Immunohistochemical localization of transforming growth factor-α ' epidermal growth factor receptor and c-erbB-2 protein in hyperplastic human prostates. Carcinogenesis *16* (1995) 729–733.
[13] Gumerlock, P. H., U. R. Poonamallee, F. J. Meyers et al.: Activated ras alleles in human carcinoma of the prostate are rare. Cancer Res. *51* (1991) 1632–1637.
[14] Ibrahim, G. K., P. E. Gravitt, K. L. Dittrich et al.: Detection of human papillomavirus in the prostate by polymerase chain reaction and in situ hybridization. J. Urol. *148* (1992) 1822–1826.
[15] Kallioniemi, A., O-P. Kallioniemi, D. Sudar et al.: Comparative genomic hybridization for molecular cytogenetic analysis of solid tumors. Science *258* (1992) 818–820.
[16] Konishi, N., Y. Hiasa, T. Tsuzuki et al.: Comparison of ras activation in prostate carcinoma in Japanese and American men. Prostate *30* (1997) 53–57.
[17] Kuhn, E. J., R. A. Kurnot, I. A. Sesterhenn et al.: Expression of the c-erbB-2 (HER-2/neu) oncoprotein in human prostatic carcinoma. J. Urol. *150* (1993) 1427–1433.
[18] Liang, P., L. Averboukh, A. B. Pardee: Distribution and cloning of eukaryotic mRNAs by means of differential display: refinements and optimization. Nucleic Acids Res. *21* (1993) 3269–3275.
[19] McNicol, P. J., J. G. Dodd: High prevalence of human papillomavirus in prostate tissues. J. Urol. *145* (1991) 850–853.
[20] Mellon, K., S. Thompson, R. G. Charlton et al.: p53, c-erbB-2 and the epidermal growth factor receptor in the benign and malignant prostate. J. Urol. *147* (1992) 496–499.
[21] Moul, J. W., P. A. Friedrichs, R. S. Lance et al.: Infrequent ras oncogene mutations in human prostate cancer. Prostate *20* (1992) 327–338.
[22] Moyret-Lalle, C., C. Marcais, J. Jacquemier et al.: ras, p53 and HPV status in benign and malignant prostate tumors. Int. J. Cancer *64* (1995) 124–129.
[23] Myers, R. B., S. Srivastava, D. K. Oelschlager et al.: Expression of $p160^{erbB\text{-}3}$ and $p185^{erbB\text{-}2}$ in prostatic intraepithelial neoplasia and prostatic adenocarcinoma. J. Natl. Cancer Inst. *86* (1994) 1140–1145.
[24] Parda, D. S., P. J. Thraves, M. R. Kuettel et al.: Neoplastic transformation of a human prostate epithelial cell line by the v-Ki-ras oncogene. Prostate *23* (1993) 91–98.
[25] Partin, A. W., J. T. Isaacs, B. Treiger et al.: Early cell motility changes associated with an increase in metastatic ability in rat prostatic cancer cells transfected with the v-Harvey-ras oncogene. Cancer Res. *48* (1988) 6050–6053.
[26] Peehl, D. M., N. Wehner, T. A. Stamey: Activated Ki-ras oncogene in human prostatic adenocarcinoma. Prostate *10* (1987) 281–289.
[27] Peehl, D. M., S. T. Wong, R. G. Sellers et al.: Loss of response to epidermal growth factor and retinoic acid accompanies transformation of human prostatic epithelial cells to tumorigenicity with v-Ki-ras. Carcinogenesis, in press.
[28] Rhim, J. S., M. M. Webber, D. Bello et al.: Stepwise immortalization and transformation of adult human prostate epithelial cells by a combination of HPV-18 and v-ki-ras. Proc. Natl. Acad. Sci. USA *91* (1994) 11874–11878.

[29] Rhim, J. S., S. Jin, M. Jung et al.: Malignant transformation of human prostate epithelial cells by N-nitroso-N-methylurea. Cancer Res., in press.

[30] Robinson, A., A. R. W. Williams, J. Piris et al.: Evaluation of a monoclonal antibody to ras peptide, RAP-5, claimed to bind preferentially to cells of infiltrating carcinomas. Br. J. Cancer *54* (1986) 877–883.

[31] Robinson, D., F. He, T. Pretlow et al.: A tyrosine kinase profile of prostate carcinoma. Proc. Natl. Acad. Sci. USA *93* (1996) 5958–5962.

[32] Ross, J. S., T. Nazeer, K. Church et al.: Contribution of HER-2/neu oncogene expression to tumor grade and DNA content analysis in the prediction of prostatic carcinoma metastasis. Cancer *72* (1994) 3020–3028.

[33] Rotola, A., P. Monini, D. di Luca et al.: Presence and physical state of HPV DNA in prostate and urinary tract tissues. Int. J. Cancer. *52* (1992) 359–365.

[34] Sadasivan, R., R. Morgan, S. Jennings et al.: Overexpression of HER-2/neu may be an indicator of poor prognosis in prostate cancer. J. Urol. *150* (1993) 126–131.

[35] Sarkar, F. H., W. A. Sakr, Y.-W. Li et al.: Detection of human papillomavirus (HPV) DNA in human prostatic tissues by polymerase chain reaction (PCR). Prostate *22* (1993) 171–180.

[36] Serfling, U., G. Ciancio, W.-Y. Zhu et al.: Human papillomavirus and Herpes virus DNA are not detected in benign and malignant prostatic tissue using the polymerase chain reaction. J. Urol. *148* (1992) 192–194.

[37] Shen, R., Z.-Z. Su, C. A. Olsson et al.: Identification of the human prostatic carcinoma oncogene PTI-1 by rapid expression cloning and differential RNA display. Proc. Natl. Acad. Sci. USA *92* (1995) 6778–6782.

[38] Shibato, M.-A., J. M. Ward, D. E. Devor et al.: Progression of prostatic intraepithelial neoplasia to invasive carcinoma in C3(1)/SV40 large T antigen transgenic mice: histopathological and molecular biological alterations. Cancer Res. *56* (1996) 4894–4903.

[39] Sun, Y., J. Lin, A. E. Katz et al.: Human prostatic carcinoma oncogene PTI-1 is expressed in human tumor cell lines and prostate carcinoma patient blood samples. Cancer Res. *57* (1997) 18–23.

[40] Suzuki, H., A. Komiya, S. Aida et al.: Detection of human papillomavirus DNA and p53 gene mutations in human prostate cancer. Prostate *28* (1996) 318–324.

[41] Terris, M. K., D. M. Peehl: Human papillomavirus detection by PCR in benign and malignant prostate tissue is dependent on the primer set utilized. Urology, in press, 1997.

[42] Thompson, T. C., J. Southgate, G. Kitchener et al.: Multistage carcinogenesis induced by ras and myc oncogenes in a reconstituted organ. Cell *56* (1989) 917–930.

[43] Tu, H., S. C. Jacobs, W. J. Mergner et al.: Rare incidence of human papillomavirus types 16 and 18 in primary and metastatic human prostate cancer. Urology *44* (1994) 726–731.

[44] Varmus, H.: An historical overview of oncogenes. In: R. A Weinberg (ed.): Oncogenes and the molecular origins of cancer, pp. 3–44. Cold Spring Harbor Laboratory Press, Cold Spring Harbor, New York 1989.

[45] Villa, L. L., R. R. Brentani: Human papillomavirus up-date. Int. J. Cancer *48* (1991) 163–166.

[46] Viola, M. V., F. Fromowitz, S. Oravez et al.: Expression of ras oncogene p21 in human prostate cancer. N. Eng. J. Med. *314* (1986) 133–137.

[47] Watanabe, M., T. Shiraishi, R. Yatani et al.: International comparison on ras gene mutations in latent prostate carcinoma. Int. J. Cancer *58* (1994) 174–178.

[48] Webber, M. M., D. Bello, S. Quader: Immortalized and tumorigenic adult human prostatic epithelial cell lines: characteristics and applications part 1. Cell markers and immortalized nontumorigenic cell lines. Prostate *29* (1996) 386–394.

[49] Weijerman, P. C., J. J. König, S. T. Wong et al.: Lipofection-mediated immortalization of human prostatic epithelial cells of normal and malignant origin using human papillomavirus type 18 DNA. Cancer Res. *54* (1994) 5579–5583.
[50] Wideroff, L., D. Schottenfeld, T. E. Carey et al.: Human papillomavirus DNA in malignant and hyperplastic prostate tissue of black and white males. Prostate *28* (1996) 117–123.
[51] Zhau, H. E., D. S. Wan, J. Zhou et al.: Expression of c-erbB-2/neu proto-oncogene in human prostatic cancer tissues and cell lines. Mol. Carcinogen. *5* (1992) 320–327.

Tumour suppressor genes in prostate cancer

S. F. Brewster

Introduction

Tumour formation results from loss of the balance between cell division and withdrawal from the cell cycle by differentiation or programmed cell death (apoptosis); invasion and metastasis, hallmarks of malignancy, result from disorders in the interactions between the cancer cells and their environment. Factors pivotal to prostatic tumourigenesis are emerging as a result of intensive multinational collaborative efforts by molecular biologists and urologists which began less than ten years ago. These comprise activation of specific cellular protooncogenes and inactivation of tumour suppressor genes. Concurrently, other systems including telomerase reactivation, production of angiogenic growth factors and carcinogen detoxifiers, androgen receptor alterations and paracrine interactions between cancer cells and prostatic stroma are under exploration. Identification of the genes frequently altered in prostate cancer is fundamental to the development of strategies enabling the screening of high-risk family members, comprehensive prognostic analysis of tumour biopsies and corrective gene therapy [9].

Tumour development has long been considered to take place in a series of steps: "... cancer is caused by mutations multiplied and accumulated usually through large-scale proliferation of cells"; "... the theory is put forward that the cancerous cell contains not one but a number of mutated genes" [50]. Accordingly, multistep epithelial carcinogenesis requires an accumulation of genetic alterations in upstaging from normal epithelium to dysplasia to *in situ* carcinoma to basement membrane, lymphatic and vascular invasion and ultimately metastasis. A molecular model for multistep colorectal carcinogenesis was developed [60, 24] which led to explorations of these observed genetic alterations as prognostic indicators (for example, ref. [36]); similar models for other sporadic cancers are currently under construction. Somatic genetic changes usually confer to the cancer cell a growth advantage over surrounding cells in terms of replicative and invasive potential: this forms the molecular basis of clonal cell population expansion seen phenotypically in primary and metastatic cancers [51]. However, malignant clones in the same primary cancer may exhibit different somatic genetic changes which result in a varying growth advantages; this may explain the histological and behavioural heterogeneity that are common features of many cancers, including prostatic carcinoma.

Tumour suppressor genes

Tumour suppressor genes (TSGs) are a diverse group of genes scattered around the normal genome and may be inactivated by a number of mechanisms. Non-random allelic loss occurs frequently in neoplasia and can be separated into three categories: first, the loss may be a primary abnormality, pathogenetically essential for development of the tumour; secondary losses responsible for phenotypic heterogeneity may accrue in later disease phases as a result of genetic instability in the cancer cells; finally random, non-clonal "background noise" loss occurs, probably not contributing to the acquisition of tumour growth advantage. It has been suggested that chromosome regions lost in less than 25% of tumours may fall into this last category [46, 55]. It is recognised from studies of retinoblastoma that somatic allelic loss at particular loci, detected as loss of heterozygosity (LOH) in allelotype studies, may indicate the presence of TSGs or their promotor regions. Rarely but significantly, both alleles are lost at a given locus, termed homozygous deletion [14]. Transcriptional silencing by DNA hypermethylation within the gene or promotor and point mutations either within gene exons or at splice donor/acceptor sites may also result in functionally inactive TSGs.

TSGs encode proteins whose functions are to maintain normal phenotype and behaviour, to promote cell–cell or cell–matrix adhesion/interaction and induction of apoptosis in the differentiated or DNA-damaged cell. Some TSGs are altered in many cancers (for example, p53), whilst others appear to be involved in only one cancer type (for example, the nephroblastoma TSG, WT-1). Because their existance only becomes apparent when inactivated, TSGs are more difficult to identify than oncogenes. Strategies include somatic allelic deletion analysis of tumour DNA followed by identification of putative genes within small frequently-deleted regions by positional cloning techniques; a second approach involves functional studies whereby phenotypic and/or behavioural suppression is observed following introduction of wild-type genetic material into cancer cells. For example, tumourigenicity is assessed by the rate of colony formation of cells in vitro and metastatic ability is assessed by the number and size of metastases developing in athymic (nude) mice. These studies can be accomplished by somatic cell hybridisation effecting transfer of an entire genome, microcell-mediated chromosome or subchromosomal transfer or single gene transfer by electroporation or viral transfection.

Allelotype studies

Two prostate cancer allelotype studies were published in the early 1990s that opened up numerous avenues for TSG research. The first study from Baltimore, USA, sampled 24 tumours obtained at radical prostatectomy for clinically-localised disease plus four obtained at transurethral resection from men with hormone-independent metastatic disease [13]. Southern blots were hybridised to seventeen polymorphic DNA probes on eleven chromosome arms. Allelic loss was observed on at least one chromosome arm in 54% of clinically localised tumours and all metastatic tumours studied. No comment was made relating LOH to tumour grade. Deletions occurred most frequently on chromosomes 10q and 16q in 29% and 31% of tumours respectively. The frequent allelic losses on 10q were expected from earlier cytogenetic reports [11], although only 10% tumours showed LOH on 7q, another cytogenetic deletion "hot-spot". Other less common deletion sites included areas harbouring candidate TSGs: chromosomes 13q14 (the retinoblastoma susceptibility gene Rb-1 [41]) 23%; 17p13.3 (near p53 at the YNZ22 locus), 17%; 18q21.1 (the Deleted in Colorectal Carcinoma gene DCC [23]), 17%. Chromosome 17p LOH was exhibited by 12% informative clinically-localised tumours and 25% advanced tumours. LOH of DCC was observed in 17% clinically-localised tumours, but the four advanced tumours sampled were not informative. The second study, from Stockholm, Sweden, involved 18 tumour samples: only two were clinically-localised, the others being locally-advanced primary tumours or samples obtained from lymph node or brain metastases [2, 39]. Blots of tumour and matched lymphocyte DNA pairs were hybridised to 54 polymorphic DNA probes mapping to 45 chromosomal arms. No abnormalities were found in 7 of the 10 primary cancers. The remainder exhibited LOH on at least one chromosome arm, those tumours being of higher histological grade. Frequent LOH was observed on 10q (30%), 16q (60%) and, supporting previous cytogenetic data, on 8p (50%). LOH at 18q22 (telomeric to DCC) was seen in 1/4 locally-advanced primary cancers and in 2/3 metastases, overall 43%. No LOH was observed on chromosomes 5q or 17p; LOH on 17q was seen in 11% samples. The discrepancies between the two studies in the percentage of tumours showing LOH, particularly on 16q and 18q are perhaps explained by the small sample sizes and sampling bias: in the Baltimore study, most were clinically-localised cancers whereas in the Swedish study, most were locally-advanced primary or metastatic tissue. Conversely, however, the percentages showing 10q LOH are identical in the two studies.

The findings of these two studies suggested the presence of TSGs on chromosomes 8p, 10q, 16q and 18q. The polymerase chain reaction (PCR) and *in situ* hybridisation techniques have now replaced Southern hybridisation for detecting allelic loss. They have the advantage that less tumour material is required, a higher percentage of tumours are informative and more detailed deletion-mapping can be obtained. Fig. 1 shows the maximum percentage allelic losses for several chromosome arms reported in the literature to date, illustrating additional regions of interest not exposed by the original allelotype studies, including chromosomes 5q and 7q. Ten areas of interest will be discussed below, believed by the author to be the most advanced in terms of research effort to date.

Chromosome 5q

LOH was observed at the APC (Adenomatous Polyposis Coli) gene locus (5q21) in 25% tumours, all of advanced stage [8]. Morton [47] reported homozygous and intragenic deletions of the E-cadherin-associated protein α-catenin, mapped to 5q31. Replacement of deleted α-catenin in the human PC-3 prostate cancer cell line suppresses tumour growth in nude mice [22]. Furthermore, reduced α-catenin mRNA expression was observed in 10/19 (53%) of primary cancers, while only 16% exhibited reduced expression of APC (T. S. Bates, unpublished). While these data indicate a tumour suppressor role for α-catenin in prostate cancer, it is noteable that APC interacts with ß-catenin and is thus involved with the same cell adhesion complex [56].

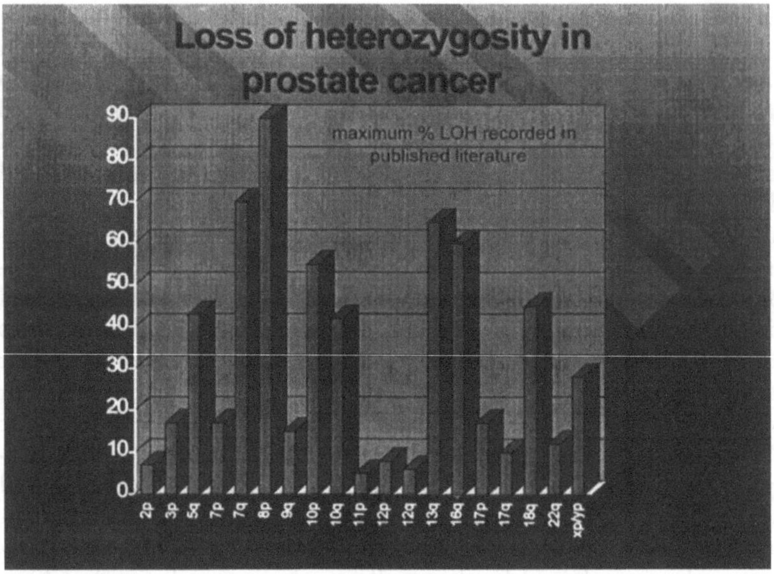

Fig. 1 A graph showing the maximum reported percentage loss of heterozygosity on 18 chromosome arms in human prostate cancer.

Chromosome 7q

Because a number of cytogenetic analyses carried out in the 1980s reported frequent deletions of portions of chromosome 7 and a murine squamous cell carcinoma is suppressed by transfer of human chromosome 7, deletion-mapping has been undertaken by two groups on tumours removed at radical prostatectomy for clinically-localised disease. Both observed maximal LOH at 7q31.1, with frequencies of 83% [62] and 30% [57]. The reason for the discrepancy between these percentages may lie in the small number of samples (sixteen) used in the first study. There was a correlation between 7q LOH and increasing tumour grade and lymph node metastasis in the second study.

Chromosome 8p

Several groups are currently investigating 8p: studying 52 primary and metastatic cancers, Bova [7] observed LOH in 63% of tumours, including one rare homozygous deletion in a lymph node metastasis. The region of commonest deletion was 8p22 adjacent to the macrophage scavenger receptor (MSR) locus. The same region is frequently deleted in colorectal and liver cancers; it contains a novel gene N33 which is hypermethylated in colon cancer cell lines, silencing its expression [6]. Functional studies of N33 in prostate cancer are awaited. 8p allelic losses were not associated with tumour grade or stage, implying that it represents an early event in prostatic carcinogenesis. However, when human chromosome 8 was transfected into a highly metastatic Dunning rat prostate cancer cell line AT6.1, metastatic ability was suppressed while in vivo growth rate and tumourigenicity were unchanged [34]. Makoska [42] has identified two additional putative TSG loci at 8p21 (near the NEFL locus) and 8p11.2 (ANK1 locus). In one of the few studies of microdissected tumours containing high-grade prostatic intraepithelial neoplasia (PIN), Emert-Buck [21] observed 8p21 LOH in 63% of PIN foci, showing intratumoural heterogeneity, while invasive components from the same samples exhibited a 90% LOH at this locus. Other candidate suppressors include the clusterin and bone morphogenetic protein 1 genes, but currently the three (or more) prostatic TSGs on 8p remain unidentified.

Chromosome 10

As discussed above, there are consistent observations of cytogenetic deletions and LOH on chromosome 10 in the literature. A deletion-mapping study has revealed LOH in up to 62% primary cancers, maximal at the q23–24 boundary, showing no correlation with tumour stage or grade [29]. Allelic loss and mutations of a candidate gene *MXI1*, located at 10q24–25, have been described in four human cancers [18]. A negative regulator of the c-*myc* oncogene, *MXI1* lies outside the most frequently deleted region. Functional studies have demonstated suppression of the metastatic rat AT6.3 subline by introduction of human chromosome 10 [49] and *in vivo* suppression of tumourigenesis with in vitro induction of apoptosis by transfer of 10pter-q11, designated *PAC1*, into human PC-3 cells [54].

Chromosome 11p

Although neither allelotype study demonstrated LOH on chromosome 11p, introduction of portions of normal human chromosome 11p into a highly-metastatic rat prostatic carcinoma cell line (AT6.1) resulted in suppression of metastatic ability without suppression of growth or tumourigenicity [33]. Chromosome 11p11.2–13, the region to which this putative metastasis suppressor gene maps, is located centromeric to WT1 (11p13) and the Beckwith-Wiedemann syndrome locus (11p15) [53]. A gene, *KAI1*, located at 11p11.2 has been cloned which is downregulated in 70% of untreated primary cancers and downregulated or absent in lymph node metastases and 90% androgen-independent primary cancers, although allelic losses and mutations have not been observed [16, 17]. *KAI1* encodes a transmembrane glycoprotein which may be involved in cell-cell or cell-matrix interactions and it appears to be a prostatic metastasis suppressor gene.

Chromosome 13q

Interest has been focused on Rb-1 at 13q14. A truncated Rb-1 transcript exhibiting a 105 nucleotide deletion was demonstrated in one out of three prostate cancer cell-lines [3]. Bookstein restored wild-type Rb-1 into the cell-line, resulting in suppression of tumourigenicity in nude mice, with unaltered growth rate in culture [4]. Seven human prostate cancers were similarly analysed, of which one (metastatic) specimen did not express Rb-1-encoded protein. Further investigation of this sample revealed LOH at the Rb-1 locus and a deletion of the transcriptional start sites from the remaining allele: a two-hit Rb-1 inactivation in an advanced prostate cancer has thus been demonstrated, proving that this TSG is inactivated in at least a subset of tumours. It has been suggested that Rb-1 alteration may occur in tumours exhibiting neuroendocrine features. Allelic loss is observed in 27–66% of primary cancers, unrelated to tumour stage or grade [10, 52].

Chromosome 16q

The gene encoding E-cadherin (E-cad), a calcium-dependent cell adhesion molecule, is located at 16q22.1. Umbas reported absent or reduced E-cad expression correlates with increasing tumour grade and stage, metastasis, progression after radical prostatectomy and reduced survival of patients with advanced prostate cancer [58, 59]. Hypermethylation of the E-cad promotor can silence expression in human samples and restoration of E-cad function by replacement of deleted α-catenin in the human PC-3 prostate cancer cell line suppresses tumour growth in nude mice [22], although *in vivo* functional studies of E-cad in prostate cancer are still awaited. Detection of a higher frequency of allelic loss telomeric to E-cad at 16q24 by comparative genomic hybridisation deletion-mapping implies a second TSG resides on 16q [15]. Candidate genes include the cell adhesion regulator (CAR) complex and H-cadherin.

Chromosome 17

The most extensively studied TSG, p53, is located at 17p13.1. Wild-type p53 appears to initiate the apoptotic cascade by negatively regulating bcl-2 and upregulating bax, a cell-death effector gene. Mutations and allelic losses displaying inter- and intratumoural heterogeneity occur in 10–22% primary tumours [19, 20, 45] and Isaacs [45] observed LOH in the adjacent pYNZ22 region (17p13.3) in two primary cancers and 2/5 cell lines. Sequencing the intact alleles revealed point mutations in three of these four specimens. A third cell line exhibited p53 mutations on both alleles. Transfection of wild-type p53 caused growth arrest in the two prostate cancer cell lines, demonstrating a suppressor role for p53 in prostatic tumourigenesis. Several immunohistochemical antibodies against p53 have become available: increased expression, presumed to represent mutant p53 product, is observed in 13–65% of primary and metastatic tumours, increasing up to 94% in androgen-independent cancers. Many studies have correlated increased p53 immunoreactivity with increasing primary tumour grade and stage, shorter time to disease progression after androgen-ablation and the androgen-independent state [5, 12, 31, 32], although McDonnell found no correlation between p53 protein accumulation in bone metastases and tumour burden or patient survival [44]. However, in examining 45 well-differentiated clinically undetected "latent" cancers for altered p53 and c-*myc* oncoprotein expression, Fox [25] concluded that neither parameter was associated with a significantly worse overall survival. All these data suggest that p53 inactivation is a late event in prostate cancer progression. Recently, Bauer [1] reported p53 and bcl-2 oncogene immunoreactivity in 175 radical prostatectomy samples for clinically-localised disease. Multivariate regression analysis showed these markers to be independent prognostic indicators of 5-year disease-free survival: if both stained positive 75% patients developed recurrent disease, while if neither marker was immunoreactive 80% remained disease-free. If confirmed by other groups, this combination immunohistochemistry could become a clinically-useful prognostic tool in the near future.

A report of LOH at 17p13.3 from a nodal metastasis but not from the primary tumour, with retention of heterozygosity at p53, has raised the question of whether inactivation of an unidentified TSG linked to the YNZ22 locus could be associated with tumour progression [20]. A TSG named HIC-1 (Hypermethylated in Cancer) spanning the YNZ22 locus has been identified and is highly expressed in benign prostate tissue [43]. HIC-1 contains a p53 binding site and a Zinc-finger DNA binding motif, suggesting that HIC-1 is a p53-responsive transcription factor. Suppression of tumourigenicity is observed when HIC-1 is transfected into breast and colon cancer cell lines, but the significance of HIC-1 in prostate cancer is currently unclear.

Chromosome 17q has attracted interest partly because the familial breast cancer gene, BRCA1, resides at 17q21 and partly because transfer of 17pter-q23 suppressed the metastatic ability of the AT6.1 cell line [53] and transfer of 17q to the human PPC-1 cell line reduced its tumourigenicity [48]. The prostatic TSG may be BRCA1 or a gene located within 2 Mb, since deletions are mapped to 17q21 in up to 44% cancers [27, 28]; it is unlikely to be nm23-H1 (located telomeric to BRCA1 at 17q21.3) since allelic loss is rarely observed at this locus [8]. Williams [61] has observed subchromosomal losses occuring telomeric to BRCA1 using fluorescence in situ hybridisation, suggesting the presence of another unidentified prostatic TSG.

Chromosome 18q

DCC is a candidate suppressor gene at 18q21.1: encoding an immunoglobulin-like cell-surface molecule that may play a role in cell-cell adhesion or interaction, this gene is deleted in 17–38% organ-confined tumours [13, 40] and a region telomeric to DCC (18q22) is deleted in 43% of locally-advanced or metastatic cancers [39]. Gao undertook a PCR-based study of DCC in 14 cancers employed primers for three regions of the gene [26]. The cancers were obtained at radical prostatectomy implying early-stage disease, although no grading or staging data were documented. 45% of the tumours showed LOH at one or other polymorphic site. Using reverse-transcriptase-PCR to amplify mRNA from the same tumours, this group found reduced DCC expression in 86% tumours including the five exhibiting LOH. Brewster [8] deletion-mapped 30 prostate cancers in which 7 out of 26 (27%) exhibited 18q LOH which significantly correlated with advanced tumour stage; one tumour retained both DCC alleles but exhibited a loss at 18q22, telomeric to DCC. These data imply the presence of a TSG on 18q, possibly telomeric to DCC, whose inactivation may confer invasive potential in prostate cancer. A second candidate TSG is DPC4 (Deleted in Pancreatic Carcinoma locus 4)], located at 18q21.1 [30]. There are no published functional data for 18q in prostate cancer.

Finally, a Texas group has reported loss of expression of an androgen-dependent immunoglobulin-like epithelial cell adhesion molecule (C-CAM) in prostate cancers [37]. Reduced C-CAM expression in benign prostatic hyperplasia and high grade PIN led to the conclusion that this change may be an early event in prostatic neogenesis. Adenoviral-mediated transfer of C-CAM into PC-3 cells suppressed tumourigenicity in nude mice and so it was proposed as a candidate for prostate cancer gene therapy [38]. The genomic location of C-CAM is not discussed.

Conclusions

In this brief review it has not been possible to discuss all the areas of interest with respect to prostatic tumour suppressor genes: there are at least twenty such areas on thirteen chromosomes. It is conceivable that each of these areas (and others) will eventually lead to the identification of genes that becomes inactivated during the development of this most complex cancer. There are five prostatic TSGs currently identified: *KAI*1, p53, E-Cadherin, Rb-1 and C-CAM. The first three of these are thought to act late in the development of the disease. There is clearly much to learn in this exciting field of research, in which we knew almost nothing before 1990.

References

[1] Bauer, J. J., I. A. Sesterhenn, F. K. Mostofi et al.: Elevated levels of apoptosis regulator proteins p53 and bcl-2 are independent prognostic biomarkers in surgically-treated clinically localized prostate cancer. J. Urol. *156* (1996) 1511–1516.

[2] Bergerheim, U. S. R., K. Kunimi, V. P. Collins et al.: Deletion mapping of chromosomes 8, 10, and 16 in human prostatic carcinoma. Genes, Chromosomes & Cancer *3* (1991) 215–220.

[3] Bookstein, R., P. Rio, S. A. Madreperla et al.: Promoter deletion and loss of retinoblastoma gene expression in human prostate carcinoma. Proc. Natl. Acad. Sci. *87* (1990) 7762–7766.

[4] Bookstein, R., J-H. Shew, P-L. Chen et al.: Suppression of tumorigenicity of human prostate carcinoma cells by replacing a mutated RB gene. Science *247* (1990) 712–715.

[5] Bookstein, R., D. MacGrogan, S. G. Hilsenbeck et al.: p53 is mutated in a subset of advanced-stage prostate cancers. Cancer Res. *53* (1993) 3369–3373.

[6] Bookstein, R., G. S. Bova, D. MacGrogan et al.: Tumour-suppressor genes in prostatic oncogenesis: a positional approach. Br. J. Urol. *79* (Suppl. 1) (1997) 28–36.

[7] Bova, G. S., B. S. Carter, M. J. Bussemakers et al.: Homozygous deletion and frequent allelic loss of chromosome 8p22 loci in human prostate cancer. Cancer Res. *53* (1993) 3869–3873.

[8] Brewster, S. F., S. Browne, K. W. Brown: Somatic allelic losses at the DCC, p53, nm23-H1 and APC tumor suppressor gene loci in prostate cancers. J. Urol. *151* (1994) 1073–1077.

[9] Brewster, S. F., J. W. Simons: Gene therapy in urological oncology: principles, strategies and potential. Eur. Urol. *25* (1994) 177–182.
[10] Brooks, J. D., G. S. Bova, W. B. Isaacs: Allelic loss of the retinoblastoma gene in primary prostatic carcinomas. The Prostate *26* (1995) 35–39.
[11] Brothman, A. R., D. M. Peehl, A. M. Patel et al.: Frequency and pattern of karyotypic abnormalities in human prostate cancer. Cancer Res. *50* (1990) 3795–3803.
[12] Byrne, R. L., C. H. Wilson-Horne, M. C. Robinson et al.: The expression of waf-1, p53 and bcl-2 in prostatic adenocarcinoma. Br. J. Urol. *79* (1997) 190–195.
[13] Carter, B. S., C. M. Ewing, W. S. Ward et al.: Allelic loss of chromosome 16q and 10q in human prostate cancer. Proc. Natl. Acad. Sci. *87* (1990) 8751–8755.
[14] Cavenee, W. K., T. P. Dryja, R. A. Phillips et al.: Expression of recessive alleles by chromosomal mechanisms in retinoblastoma. Nature *305* (1983) 775–784.
[15] Cher, M. L., T. Ito, N. Weidner et al.: Mapping of regions of physical deletion on chromosome 16q in prostate cancer cells by fluorescence *in situ* hybridization (FISH). J. Urol. *153* (1995) 249–254.
[16] Dong, J.-T., P. W. Lamb, C. W. Rinker-Schaeffer et al.: *KAI1*, a metastasis suppressor gene for prostate cancer on human chromosome 11p11.2. Science *268* (1995) 884–886.
[17] Dong, J-T., H. Suzuki, S. S. Pin et al.: Down-regulation of the *KAI1* metastasis suppressor gene during the progression of human prostatic cancer infrequently involves gene mutation or allelic loss. Cancer Res. *56* (1996) 4387–4390.
[18] Eagle, L. R., X. Yin, A. R. Brothman et al.: Mutation of the *MXI1* gene in prostate cancer. Nature Genet. *9* (1995) 249–255.
[19] Effert, P. J., A. Neubauer, P. J. Walther et al.: Alterations of the p53 gene are associated with the progression of a human prostate carcinoma. J. Urol. *147* (1992) 789–793.
[20] Effert, P. J., R. H. McCoy, P. J. Walther et al.: p53 gene alterations in human prostate cancer. J. Urol. *150* (1993) 257–261.
[21] Emert-Buck, M. R., C. D. Vocke, R. O. Pozzatti et al.: Allelic loss on chromosome 8p12–21 in microdissected prostatic intraepithelial neoplasia. Cancer Res. *55* (1995) 2959–2962.
[22] Ewing, C. M., N. Ru, R. A. Morton et al.: Chromosome *5* suppresses tumorigenicity of human PC-3 prostate cancer cells- correlation withre-expression of a-catenin and restoration of E-Cadherin function. Cancer Res. *55* (1995) 4813–4816.
[23] Fearon, E. R., K. R. Cho, J. M. Nigro et al.: Identification of a chromosome 18q gene that is altered in colorectal cancer. Science *247* (1990) 49–55.
[24] Fearon, E. R., B. Vogelstein: A genetic model for colorectal tumorigenesis. Cell *61* (1990) 759–767.
[25] Fox, S. B., R. A. Persad, J. Royds et al.: p53 and c-*myc* expression in stage A1 prostatic adenocarcinoma: useful prognostic determinants? J. Urol. *150* (1993) 490–494.
[26] Gao, X., K. V. Honn, D. Grignon et al.: Frequent loss of expression and loss of heterozygosity of the putitive tumor suppressor gene DCC in prostate cancers. Cancer Res. *53* (1993) 2723–2727.
[27] Gao, X., A. Zacharek, A. Salkowski et al.: Loss of heterozygosity of the BRCA1 and other loci on chromosome 17q in human prostate cancer. Cancer Res. *55* (1995) 1002–1005.
[28] Gao, X., A. Zacharek, D. J. Grignon et al.: Localization of potential tumor suppressor loci to a <2 Mb region on chromosome 17q in human prostate cancer. Oncogene *11* (1995) 1241–1247.
[29] Gray, I. C., S. M. Phillips, S. J. Lee et al.: Loss of the chromosomal region 10q23–25 in prostate cancer. Cancer Res. *55* (1995) 4800–4803.
[30] Hahn, S. A., M. Schutte, A. T. Hoque et al.: DPC4, a candidate tumor suppressor gene at human chromosome 18q21.1. Science *271* (1996) 350–353.

[31] Hamdy, F. C. W. Thurrell, J. Lawry et al.: p53 mutant expression correlates with hormone sensitivity and prognosis in prostatic adenocarcinoma. J. Urol. *149* (1993) 377A.
[32] Heidenberg, H. B., I. A. Sesterhenn, J. P. Gaddipati et al.: Alteration of the tumor suppressor gene p53 in a high fraction of hormone refractory prostate cancer. J. Urol. *154* (1995) 414–421.
[33] Ichikawa, T., Y. Ichikawa, J. T. Dong et al.: Localisation of metastasis suppressor gene(s) for prostatic cancer to the short arm of chromosome 11. Cancer Res. *52* (1992) 3486–3490.
[34] Ichikawa, T., N. Nihei, H. Suzuki et al.: Suppression of metastasis of rat prostatic cancer by introducing human chromosome 8. Cancer Res. *54* (1994) 2299–2302.
[35] Isaacs, W. B., B. S Carter, C. M. Ewing et al.: Wild-type p53 suppresses growth of human prostate cancer cells containing mutant p53 alleles. Cancer Res. *51* (1991) 4716–4720.
[36] Jen, J., H. Kim, S. Piantadosi et al.: Allelic loss of chromosome 18q and prognosis in colorectal cancer. N. Eng. J. Med. *331* (1994) 213–221.
[37] Kleinerman D. I., P. Troncoso, S-H. Lin et al.: Consistent expression of an epithelial cell adhesion molecule (C-CAM) during human prostate development and loss of expression in prostate cancer: implication as a tumor suppressor. Cancer Res. *55* (1995) 1215–1220.
[38] Kleinerman, D. I., W-W. Zhang, S-H. Lin et al.: Application of a tumor suppressor (C-CAM1)-expressing recombinant adenovirus in androgen-independent human prostate cancer therapy: a preclinical study. Cancer Res. *55* (1995) 2831–2836.
[39] Kunimi, K., U. S. Bergerheim, I-L. Larsson et al.: Allelotyping of human prostatic adenocarcinoma. Genomics *11* (1991) 530–536.
[40] Latil, A., J-C. Baron, O. Cussenot et al.: Genetic alterations in localised prostate cancer: identification of a common region of deletion on chromosome arm 18q. Genes, Chromosomes & Cancer *11* (1994) 119–125.
[41] Lee, W-H., R Bookstein, F. Hong et al.: Human retinoblastoma susceptibility gene: Cloning, identification and sequencing. Science *235* (1987) 1394–1399.
[42] Macoska, J. A., T. M. Trybus, P. D. Benson et al.: Evidence for three tumor suppressor gene loci on chromosome 8p in human prostate cancer. Cancer Res. *55* (1995) 5390–5395.
[43] Makos-Wales, M., M. A. Biel, W. E. Deiry et al.: p53 activates expression of HIC-1, a new candidate tumour suppressor gene on 17p13.3. Nature Med. *1* (1995) 570–577.
[44] McDonnell, T. J., N. M. Navone, P. Troncoso et al.: Expression of bcl-2 oncoprotein and p53 protein accumulation in bone marrow metastases of androgen-independent prostate cancer. J. Urol. *157* (1997) 569–574.
[45] Mirchandani, D., J. Zheng, G. J. Miller et al.: Heterogeneity in intratumoral distribution of p53 mutations in human prostate cancer. Am. J. Pathol. *147* (1995) 92–101.
[46] Mitelman, F., Y. Kaneko, J. Trent, et al.: Report of the committee on chromosome changes in neoplasia. Cytogenet. Cell Genet. *58* (1991) 1053–1079.
[47] Morton, R. A., C. M. Ewing, J. J. Watkins et al.: Reduction of E-cadherin levels and deletion of the a-catenin gene in human prostate cancer cells. Cancer Res. *53* (1994) 3585–3590.
[48] Murakami, Y. S., A. R. Brothman, R. L. Leach et al.: Suppression of malignant phenotype in a human prostate cancer cell line by fragments of normal chromosomal region 17q. Cancer Res. *55* (1995) 3389–3394.
[49] Nihei, N., T. Ichikawa, Y. Kawana et al.: Localization of metastasis suppressor gene(s) for rat prostatic cancer to the long arm of chromosome 10. Genes, Chromosomes & Cancer *14* (1995) 112–119.
[50] Nordling, C. O.: A new theory on the cancer-inducing mechanism. Br. J. Cancer *7* (1953) 68–72.
[51] Nowell, P. C.: The clonal evolution of tumor cell populations. Science *194* (1976) 23–28.

[52] Phillips, S. M., D. G. Morton, S. J. Lee et al.: Loss of heterozygosity of the retinoblastoma and adenomatous polyposis coli gene loci and in chromosome 10p, 10q and 16q in prostate cancer. Br. J. Urol. *73* (1994) 390–395.
[53] Rinker-Schaeffer, C. W., A. L. Hawkins, N. Ru et al.: Differential suppression of mammary and prostate cancer metastasis by human chromosomes 17 and 11. Cancer Res. *54* (1994) 6249–6256.
[54] Sanchez, Y., M. Lovell, M. C. Marin et al.: Tumor suppression and apoptosis of human prostate carcinoma mediated by a genetic locus within human chromosome 10pter-q11. Proc. Natl. Acad. Sci. *93* (1996) 2551–2556.
[55] Seizinger, B. R., H. P. Klinger, C. Junien et al.: Report of the committee on chromosome and gene loss in human neoplasia. Cytogenet. Cell Genet. *58* (1991) 1080–1096.
[56] Su, L. K., B. Vogelstein, K. W. Kinzler: Association of the APC tumor suppressor protein with catenins. Science *262* (1993) 1734–1737.
[57] Takahashi, S., A. L. Shan, S. R. Ritland et al.: Frequent loss of heterozygosity at 7q31.1 in primary prostate cancer is associated with tumor aggressiveness and progression. Cancer Res. *55* (1995) 4114–4119.
[58] Umbas, R., J. A. Schalken, T. W. Aalders et al.: Expression of the cellular adhesion molecule E-Cadherin is reduced or absent in high-grade prostate cancer. Cancer Res. *52* (1992) 5104–5109.
[59] Umbas, R., W. B. Isaacs, P. P. Bringuier et al.: Aberrant E-cadherin expression is correlated with poor prognosis in early and advanced stage prostate cancer. Cancer Res. *54* (1994) 3929–3933.
[60] Vogelstein, B., E. R. Fearon, S. R. Hamilton et al.: Genetic alterations during colorectal tumor-development. N. Eng. J. Med. *319* (1988) 525–532.
[61] Williams, B. J., E. Jones, X. L. Zhu et al.: Evidence for a tumor suppressor gene distal to BRCA1 in prostate cancer. J. Urol. *155* (1996) 720–725.
[62] Zenklusen, J. C., J. C. Thompson, P. Troncoso et al.: Loss of heterozygosity in human primary prostate carcinomas: a possible tumor suppressor at 7q31.1.

Appendix

Since the time of writing, a TSG on chromosome 10 (10q23) named PTEN/MMAC1 was identified in several human cancers. Mutations and allelic loss have been observed in 3 prostate cancer cell lines and 28% of primary clinically-confined tumours [Cairns et al. Cancer Res. 57 (1997) 4997]. The gene product shows sequence homology to cytoskeletal proteins, tensin and auxilin. Regarding p53, we have undertaken a study which confirms the findings of Bauer et al. (1996) and in addition we observed aberrant p53 immunohistochemistry biopsies to be an independent predictor of biochemical relapse after readical prostatectomy [Brewster et al. (1998), unpublished].

The use of differential display PCR for the detection of new tumor suppressor genes in prostate cancer

Ch. P. Pilarsky

Prostate carcinoma (CaP) is the most common internal cancer of men and the second most common cause of cancer deaths of males in the USA. The number of newly diagnosed cases has increased over the past decade. Approximately 300,000 new cases are expected to be diagnosed and roughly 38,000 deaths will occur as a result of this disease in 1997 [16]. It is commonly accepted that mutations in distinct genes are responsible for tumor development. These mutations result in the functional inactivation of these genes and lead to the corresponding inactive proteins. At the end, these events change the phenotype of the tissue from benign to neoplastic and the tumor will arise [8]. These genetic alterations that occur during carcinogenesis are mainly analysed by investigating genomic DNA for structural aberrations by various techniques. These techniques have been grouped by the minimal resolution of aberration they can identify. Cytogenetic analysis is able to analyse only gross divergences like numerical or large structural changes. Another technique for investigating DNA alterations is comparative genomic hybridisation which shares the possibility to analyse the genome at once with the cytogenetic analysis. On the other hand, analysing allelic loss in tumors, another common technique, has a resolution of < 1 cM but it is nearly impossible to analyse the total human genome at the gene level with this technique [3]. However, using these techniques and performing positional cloning for analysing the underlying events during tumor formation might result in the isolation of tumorsuppressor- and oncogenes.

All mutations in the DNA will result in the transcription of mutated RNA which therefore is reflecting the altered DNA. It is known that missense and/or nonsense mutations of a specific gene might result in changed turnover of the corresponding mRNA. Since most of the tumors are metabolically more active than their corresponding tissue the turnover rate of mRNA and proteins might be enhanced in general resulting in decreased mRNA level which leads to a decreased level of the associated protein. Stop codon mutations are commonly associated with a shortened transcript, and deletions in the gene will raise nonfunctional protein. Furthermore tumorsuppressor- and oncogenes are often part of the transcriptional machinery which therefore will lead to a direct change of expression of their controlled genes (Fig. 1).

All modern techniques of molecular genetics could be applied to RNA. Beside the possibility to analyse the RNA for the identification of tumorsuppressor- or oncogenes, there is another point which makes RNA analyses important: The analysis of quantitative changes in the total mRNA in the tumor cells will lead to a better understanding of how other genes are involved in tumor formation which are not mutated but changed in their expression level (Class II genes) [17]. These findings

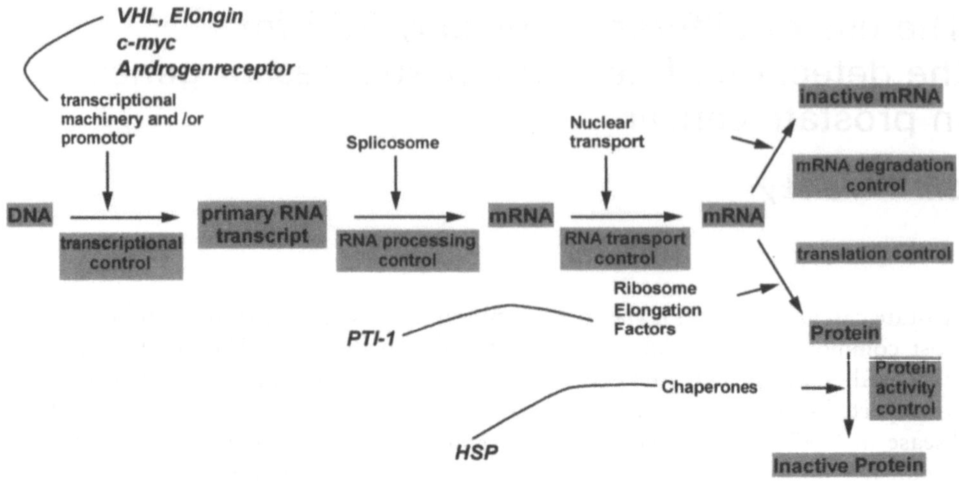

Fig. 1 The way from the DNA to protein and the different checkpoints which are regulating gene expression. There are several genes which have been involved in regulating gene expression and which are mutated in different carcinomas (indicated in italics).

might lead to the identification of distinct tumor genotypes which are normally summed up as one tumor phenotype. One type of adenocarcinoma of the prostate might therefore not only be identified by the phenotype, but it might also be characterized as prostate cancer due to mutation in Chromosome 13q and another might be identified as tumor resulting from mutations in 8p and this genotype might have an other prognosis than the former mentioned genotype. Interestingly, in bladder cancer this is possible today, since the occurence of p53 mutations is an important factor for prognosis regardless the histological type of bladder tumor.

There are several techniques to detect the quantitative changes of protein expression during tumorigenesis. If these changes are analysed on the protein level, 2-D Gelelectrophoresis might be a suitable technique. But this method is limited by the rela-

Fig. 2 Outline of the differential display PCR method. At first the crude RNA is extracted from the tissue and then reverse-transcribed using anchored primer. These anchored primer lead to a reverse transcription which starts at the poly-A-end. The cDNA is then amplified by PCR with the anchored primer used during RT and an arbitrary primer. The arbitrary primer is composed ideally of a 50% G:C content. A fraction of the PCR is subjected to a polyacrylamide gelelectrophoresis and the amplified fragments are usually detected by autoradiography. Differentially expressed cDNA fragments are eluted from the gel and reamplified using the same primer combination. This PCR product is subjected to agarose gelelectrophoresis eluted from the gel and cloned in a T/A type cloning vector. Commonly a DD-PCR band is composed of three different cDNA sequences and one of these fragments raises approximately 60% of the detected signal. Since only this fragment is used for further investigation it is identified by dot blot of individual clones of the eluted DD-PCR fragment and hybridisation with the original DD-PCR. The identified fragment is then analysed by commonly used molecular techniques as sequencing, Northern blot or library screening.

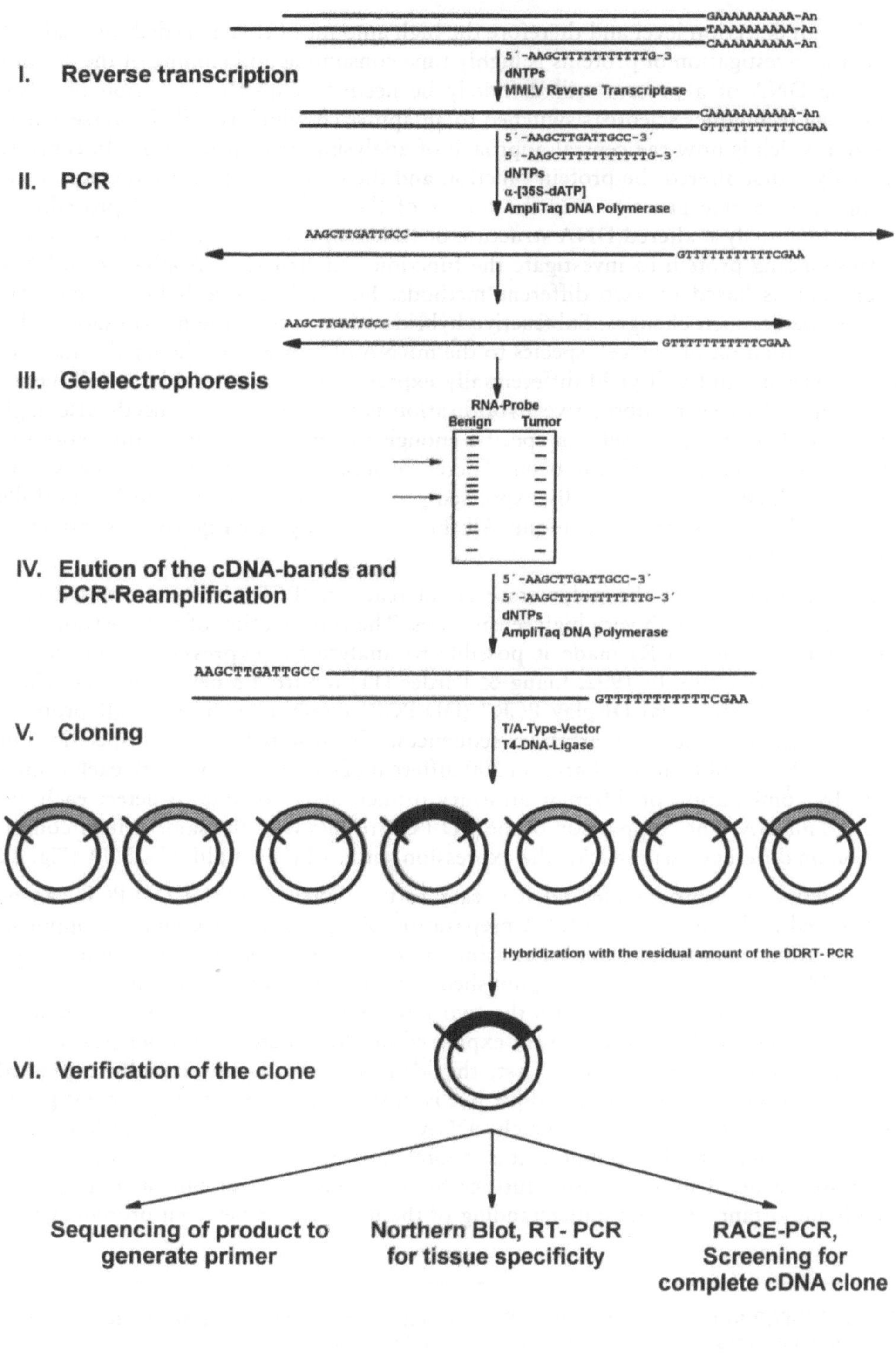
GAAAAAAAAAA-An
TAAAAAAAAAA-An
CAAAAAAAAAA-An
I. Reverse transcription
5´-AAGCTTTTTTTTTTTG-3
dNTPs
MMLV Reverse Transcriptase
CAAAAAAAAAAA-An
GTTTTTTTTTTTCGAA
II. PCR
5´-AAGCTTGATTGCC-3´
5´-AAGCTTTTTTTTTTTG-3´
dNTPs
α-[35S-dATP]
AmpliTaq DNA Polymerase
AAGCTTGATTGCC
GTTTTTTTTTTTCGAA
AAGCTTGATTGCC
GTTTTTTTTTTTCGAA
III. Gelelectrophoresis
RNA-Probe
Benign Tumor
IV. Elution of the cDNA-bands and PCR-Reamplification
5´-AAGCTTGATTGCC-3´
5´-AAGCTTTTTTTTTTTG-3´
dNTPs
AmpliTaq DNA Polymerase
AAGCTTGATTGCC
GTTTTTTTTTTTTCGAA
V. Cloning
T/A-Type-Vector
T4-DNA-Ligase
Hybridization with the residual amount of the DDRT- PCR
VI. Verification of the clone
Sequencing of product to generate primer
Northern Blot, RT- PCR for tissue specificity
RACE-PCR, Screening for complete cDNA clone

tively high detection level and therefore the high amount of tissue needed for analysis. Also the investigation of proteins is highly time consuming and cloning of the accompanying DNA of a protein will definitely be needed. With the invention of DNA analysing techniques scientists switched to an approach which is called reverse genetics and which is now the central approach of analysing protein function. In contrast to analyse first altered the protein function and then trying to identify the accompanying gene reverse genetics uses the power of the molecularbiological procedures, which first analyse altered DNA structure or RNA expression and then looks for the corresponding protein to investigate the function and structure. Analysis of mRNA expression is based on two different methods. First subtractive hybridisation was used to detect such changes. Subtractive hybridisation implies the hybridisation of a cDNA population of one cell species to the mRNA of another. In theory this method is very elegant and will yield differentially expressed genes preferably as full length transcripts. However, subtractive hybridisation is time consuming, needs also high amounts of tissue and is seldom specific enough to generate differentially expressed genes in one round of hybridisation. Therefore subtractive hybridisation leads only to an enrichment of differentially expressed genes. Furthermore, high technical skills are needed to perform this technique. All this lead to only rare reports on subtractive hybridisation [21].

With the invention of the polymerase chain reaction (PCR) it became possible to detect nucleic acids with very high sensitivities. The combination of reverse transcription and PCR (RT-PCR) made it possible to analyse the expression of mRNA in tissues and cells, too. In 1992, Liang & Pardee [11] reported a new technique which they called "Differential Display PCR" (DD-PCR) combining the RT-PCR protocol with the use of primer with arbitrary sequences. This resulted in the "unspecific" but reproducible amplification of around 200 different cDNA fragments with each primer pair. By combination of different arbitrary primers it is possible to detect each expressed mRNA. The comparison of the DD-PCR results with the same primer combination on different tissues yields the expression status of the amplified cDNA (Fig. 2).

This method is reproducible and it is easy to use. Furthermore, a DD-PCR can be performed with 3 µg of crude RNA preparation of a given tissue giving the opportunity of using small tissue samples as a fine needle biopsy. Since the invention of DD-PCR, 286 different articles were published concerning this technique indicating the reliability of this method. Besides the two key benefits, the use of low amounts of tissue and the amplification of every expressed mRNA, there are further possibilities which are offered by DD-PCR. First, the identification of differentially expressed genes is not inherited in the technique but is in the responsibility of the investigator. So the investigator can decide which cDNA fragment he accounts as differentially expressed. And with this technique it is theoretically possible to find any differentially expressed gene. This might give further hints of tumor development and appears highly important since the understanding of the interactions between proteins which

Fig. 3 DD-PCR from 4 patients with CaP; N: benign prostate tissue, T: prostate tumor tissue, the arrow indicates the fragment that was analysed in detail.

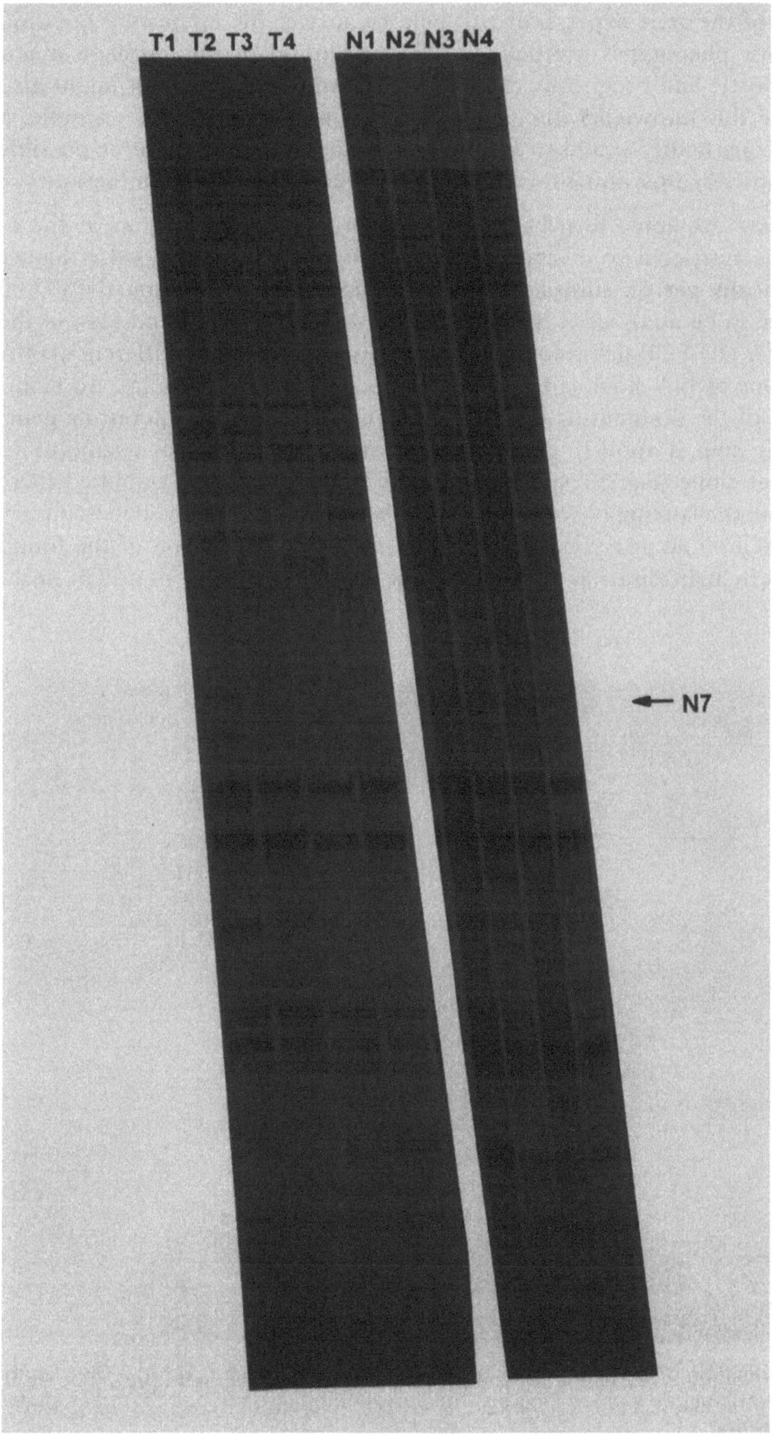
T1 T2 T3 T4
N1 N2 N3 N4
N7

result in changes of the gene expression will help to answer the interesting question how specific tumor phenotypes are defined. The resolution of this problem might lead to new diagnostic and prognostic tools in the treatment of cancer. It might also be possible to use this knowledge for new treatment modalities. If, for example, a group of tumors consistently express a special membrane protein it might be possible to develop antibodies against this protein for an antibody based immunotherapy.

Unfortunately, there are some drawbacks of DD-PCR. The amplification of the c-DNA fragments is restricted to 3′-ends of the cDNA and the work load is higher than in a theoretically perfect subtractive hybridisation since approximately 30.000 cDNA bands have to be analysed in the first round. Furthermore a band seen in the autoradiogram of a DD-PCR gel is composed of approximately three different cDNA clones, but only one cDNA clone is the major component of this band [5]. To facilitate the detection of the dominating cDNA clone, and therefore the interesting gene, a second screening step is applied. This screening step uses a dot blot technique to identify the correct clone (Fig. 2.). Since commonly only 3′-ends are amplified there is the need to clone the lacking part of the cDNA. This can be done by cDNA library screening and RACE-PCR. It is also necessary to check the expression of the found cDNA by Northern hybridisation to confirm the expression by a non-PCR technique (Fig. 2).

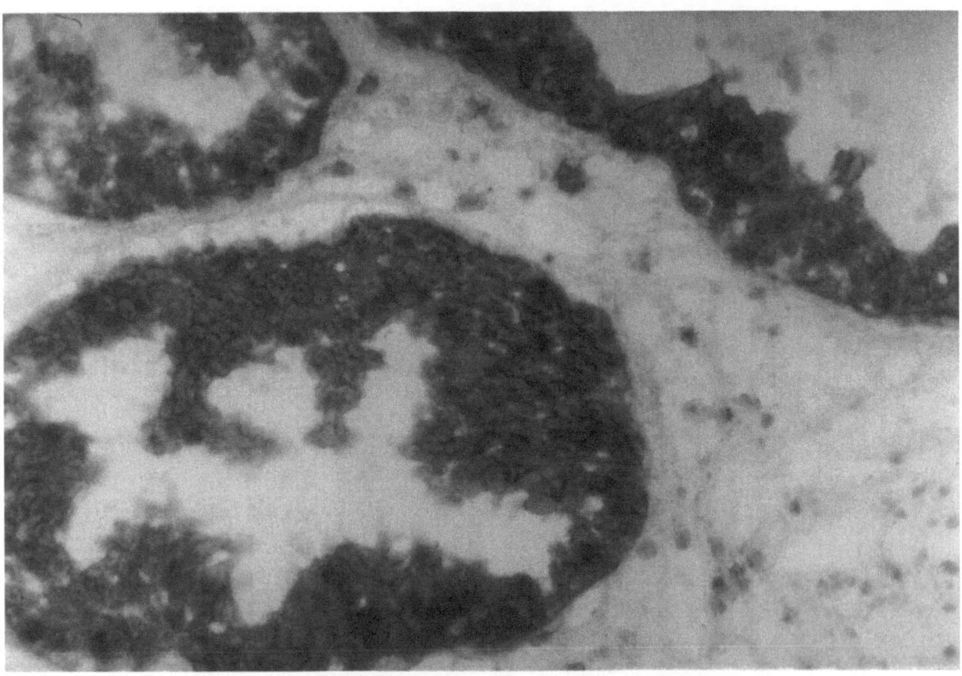

Fig. 4 In-situ-hybridisation of (a) benign prostate tissue from one patient with the Cyr61 fragment. The epithelium is highly stained whereas the surrounding stromal compartment shows only neglectable staining.

We therefore applied DD-PCR for the detection of genes whose expression is changed during CaP development. The RNA of 4 patients with untreated CaP was analysed for differentially expressed genes. With DD-PCR we identified a downregulated cDNA-fragment in prostate carcinoma cells (Fig. 3). This fragment (N7) was cloned and further analysed by comparative multiplex RT-PCR (CMRT-PCR) and in-situ-hybridisation. Sequence analysis revealed that N7 is identical to the recently described extra cellular matrix signaling protein Cyr61. By in-situ hybridisation we could show that the expression of Cyr61 is restricted to the epithelium of the prostate (Fig. 4).

CMRT-PCR with sequence specific and G3PDH primers was performed for 25 cycles with pairs of tumor and benign tissues from 13 patients. CMRT-PCR demonstrated that Cyr61 is downregulated in the tumor tissue of 7 out of 13 patients (Fig. 5).

Cyr61 is known to promote the growth of cells by the modulation of cell attachment and enhancement of growth factor effects. We could demonstrate that Cyr61 is exclusively expressed in the epithelia of the prostate. Furthermore, we were able to show that Cyr61 is downregulated in a subset of patients during prostate carcinogenesis. Regarding the functions of Cyr61, the changed expression pattern of Cyr61 might therefore attribute to altered interactions between epithelial and stromal cells in prostate cancer which could lead to growth suppression of the stromal cells. Further

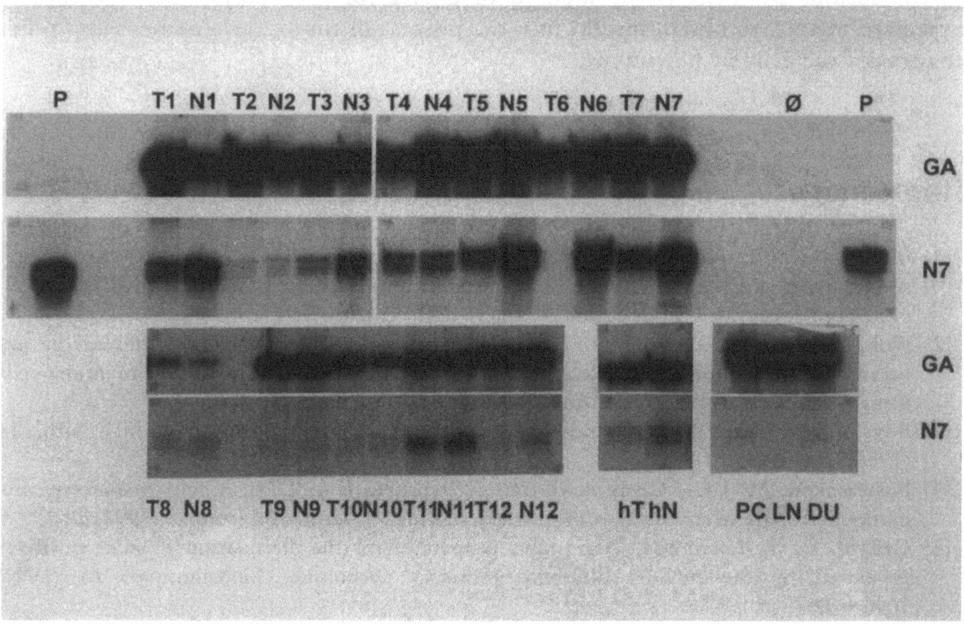

Fig. 5 Autoradiogram of a polyacrylamide gel with a comparative CMRT-PCR of Cyr61 from 12 patients with untreated CaP, one patient with neoadjuvant hormonal treatment and three different cell lines; T (prostate tumor tissue), N (benign prostate tissue): RNA from the patients with untreated CaP; hT, hN: RNA from the patient with hormonal therapy; LN: LNCaP, DU: DU145, PC: PC3; GA: GAPDH which was amplified simultaneously and served as an internal control; N7: The Cyr61 fragment which was amplified from the indicated tissue, P: size control for the PCR using the Cyr61 cDNA-fragment as template, Ø: negative control of the RT-PCR.

analyses of Cyr61 expression in prostate and other tumors might lead to a better understanding of the involvement of Cyr61 during tumorigenesis.

Our results in combination with others (Table 1) show that DD-PCR is able to detect differential expressed genes which were normally not associated with changes in the DNA (Table 1).

Table 1 References with results of differential display regarding prostate cancer

Expression changed	Identified gene	Reference
Upregulated	Macrophage migration inhibitory factor in metastases	Meyer-Siegler et al., Urology, 1996
	Early growth response-1	Thigpen et al., J. Urol. 1996
	DD3	Bussemakers et al., Urol. Res., 1997
	Cytochrome C Oxidase, COSVIc	Wang et al., Cancer Res., 1996
Downregulated	Epithelial Tropomyosin, TMe1	ibid.

In conclusion, DD-PCR is the method of choice for the analysis of differentially expressed mRNA to obtain insight into the process of tumor development and other changes of the cellular phenotype.

References

[1] Blok, L. J., M. E. Grossmann, J. E. Perry et al.: Characterization of an early growth response gene, which encodes a zinc finger transcription factor, potentially involved in cell cycle regulation. Mol. Endocrinol. *9* (1995) 1610–1620.

[2] Blok, L. J., M. V. Kumar, D. J. Tindall: Isolation of cDNAs that are differentially expressed between androgen-dependent and androgen-independent prostate carcinoma cells using differential display PCR. Anal. Biochem. *226* (1995) 383–386.

[3] Bova, G., W. Isaacs: Review of allelic loss and gain in prostate cancer. World J. Urol. *14* (1996) 338–346.

[4] Bussemakers, M. J. G., A. van Bokhoven, F. Debruyne et al: DD3: A new prostate-specific marker, strongly overexpressed in prostatic tumors. J. Urol. *157* (suppl.) (1997) 21.

[5] Callard, D., B. Lescure, L. Mazzolini: A method for the elimination of false positives generated by the mRNA differential display technique. Biotechniques *16* (1994) 1096–1097.

[6] Ikonomov, O. C., M. H. Jacob: Differential display protocol with selected primers that preferentially isolates mRNAs of moderate- to low-abundance in a microscopic system. Biotechniques *20* (1996) 1030–1034.

[7] Ito, T., N. Adati, H. Hagiwara et al.: Fluorescent differential display: arbitrarily primed RT-PCR fingerprinting on an automated DNA sequencer. FEBS Lett. *351* (1994) 231–236.

[8] Knudson, A. G.: Antioncogenes and human cancer. Proc. Natl. Acad. Sci. USA *90* (1993) 10914–10921.

[9] Liang, P., L. Averboukh, A. B. Pardee: Distribution and cloning of eukaryotic mRNAs by means of differential display: refinements and optimization. Nucleic Acids Res. *21* (1993) 3269–3275.

[10] Liang, P., Z. Guo, L. Averboukh et al.: Differential display using one-base anchored oligo-dT primers. Nucleic Acids Res. *22* (1994) 5763–5764.

[11] Liang, P., A. B. Pardee: Differential display of eukaryotic messenger RNA by means of the polymerase chain reaction. Science *257* (1992) 967–970.

[12] Liang, P., A. B. Pardee: Recent advances in differential display. Curr. Opin. Immunol. *7* (1995) 274–280.

[13] Loda, M., P. Capodieci, C. Corless et al.: Expression of mitogen-activated protein kinase phosphatase-1 in the early phases of human epithelial carcinogenesis. Am. J. Pathol. *149* (1996) 1553–1564.

[14] Magigalluzzi, C., M. Loda: Molecular events in the early phases of prostate carcinogenesis. Eur. Urol. *30* (1996) 167–176.

[15] Meyer-Siegler, K., P. B. Hudson: Enhanced expression of macrophage migration inhibitory factor in prostatic adenocarcinoma metastases. Urology *48* (1996) 448–452.

[16] Parker, S. L., T. S. B. Tong, P. A. Wingo: Cancer Statistics, 1997. CA Cancer J. Clin. *47* (1997) 5–27.

[17] Sager, R.: Expression genetics in cancer: Shifting the focus from DNA to RNA. Proc. Natl. Acad. Sci. USA *94* (1997) 952–955.

[18] Sompayrac, L., S. Jane, T. C. Burn et al.: Overcoming limitations of the mRNA differential display technique. Nucleic Acids Res. *23* (1995) 4738–4739.

[19] Thigpen, A. E., K. M. Cala, J. M. Guileyardo et al.: Increased expression of early growth response-1 messenger ribonucleic acid in prostatic adenocarcinoma. J. Urol. *155* (1996) 975–981.

[20] Wang, F. L., Y. Wang, W. Wong et al.: Two differentially expressed genes in normal human prostate tissue and in carcinoma. Cancer Res. *56* (1996) 3634–3637.

[21] Watson, J. B., J. E. Margulies: Differential cDNA screening strategies to identify novel stage-specific proteins in the developing mammalian brain. Dev. Neurosci. *15* (1993) 77–86.

Molecular alterations associated with prostate cancer development

M. J. G. Bussemakers, J. A. Schalken

1. Clinical background

Prostate cancer, the malignancy with the highest incidence rate in the western male population, is characterized by the fact that, at clinical presentation, the majority of patients have disease extended beyond the prostate, i. e., capsular penetration and/or established regional or distant metastases. The clinical consequences of established metastatic disease are profound since no curative therapy is available. Systemic palliative methods, based on androgen ablation, are usually successful, but of limited duration. The outgrowth of androgen-insensitive cells is inevitable and will eventually result in the patient's death.

One of the important questions to be addressed in the field of prostate cancer research is: what are the molecular events responsible for the development and progression of prostate cancer? Or, in other words, why and how does prostate cancer evolve from an indolent to a life-threatening disease? Moreover, why are some prostatic tumors destined to never progress and why will other prostate cancers inevitably lead to advanded disease? At this moment, no accepted method is available to make these distinctions, which makes systematic screening programs at least questionable. Therefore, methods for prediction of the metastatic potential of tumor cells are urgently needed. Many research laboratories are using molecular approaches to this problem. These approaches are all based on the question of whether molecular differences at the DNA, RNA, or protein level can be identified that can serve as diagnostic and/or prognostic markers for malignant prostatic disease. Here, an overview is given of results obtained by studies aimed at a better understanding of the molecular basis of prostate cancer development and at the identification of prognostic markers for this disease.

2. Molecular changes in prostate cancer

Tumor progression is associated with phenotypic changes, which are determined by the pattern of protein expression. Aberrant protein expression patterns that mark the progression, however, often find their origin in genomic alterations. Therefore, studies on the involvement of specific genetic changes, such as chromosomal abnormalities, can serve as a lead to identify genes that are implicated in the onset and progression of cancer. The molecular biology of human cancer has seen explosive progress in the last decade. Several cancer susceptibility genes have been identified and have provided a basis for the understanding of the relationship between inherited malignancies and sporadic forms of common human cancers. However, despite the high prevalence of prostate cancer, little is known with regard to the molecular pathogenesis of this disease.

2.1 Prostate cancer susceptibility genes

The initiation of prostate cancer, i. e., the formation of a histologically identifiable lesion, appears to be a very frequent event. Geographically, this rate of histological cancer incidence is roughly the same world-wide [75]. However, the number of clinically manifest cases differs widely among various populations, suggesting an important role for environmental factors in triggering this progression. Furthermore, studies of familial aggregation of this disease have suggested that between 5–10% of prostate cancers may be caused by a prostate cancer susceptibility gene which may act as a genetic factor driving progression of the disease independent of environmental exposure. As with numerous other cancers, there is evidence for both genetic and environmental factors involved in the etiology of prostate cancer, with the majority of disease most likely being a result of the interaction between the two.

Familial clustering of prostate cancer has already been described in 1960 and several subsequent studies have confirmed this observation. Extensive cancer pedigrees have been studied and segregation analysis showed that the familial clustering did not reflect a shared environmental risk factor but was due to a genetic mechanism. Familial clustering was indeed best explained by dominant inheritance of a rare high-risk allele [11]. Linkage analysis in families with multiple members affected with prostate cancer are currently being used to identify the prostate cancer susceptibility gene(s). It is anticipated that these studies will provide critical information regarding the genetic mechanisms not only for hereditary prostate cancer but also of the more common sporadic disease as well. Recently, a major susceptibility locus for prostate cancer on chromosome 1 (1 q24–25) was suggested by a genome wide search, however, the actual gene(s) involved still remain to be identified [52].

2.2 Chromosomal aberrations in prostate cancer

Unlike other solid tumors, cytogenetic analyses of prostate cancer specimens have revealed few consistent chromosomal deletions. More recently, techniques have become available to study chromosomal losses at the molecular level, like Restriction Fragment Length Polymorphism (RFLP) analysis, microsatellite PCR, and comparative genomic hybridization (CGH). Using these techniques a number of chromosomal regions have been identified that are frequently affected in prostate cancer. Allelic loss of chromosome 8 p has been reported by several studies [7, 40, 41, 65, 71] and research is ongoing to identify and characterize the putative tumor suppressor gene, located on chromosome 8 p, which is involved in prostate cancer development. Interestingly, chromosome 8 p is also frequently lost in colorectal cancer (45%), non-small cell lung cancer (50%), hepatocellular carcinoma (45%) and bladder cancer (25%).

A number of candidate genes in the affected region have been identified, including a homeobox gene [25]. However, more studies are required to further identify which gene is affected in prostate cancer and whether the same gene is affected in all the other cancers as well. Besides frequent loss of 8 p, gain of chromosome 8 q has been observed in a subset of prostate tumors [7, 13, 41] suggesting the presence of an activated oncogene on this chromosome as well. A candidate gene which may be the target of these amplification events is the c-myc proto-oncogene located at 8 q24. Overexpression of c-myc has been reported in prostate cancer, however, additional studies are necessary to establish the correlation between gain of chromosome 8 q and c-myc (over)expression.

Other chromosomal regions that are frequently lost in prostate cancer are 16 q (30%) and 5 q (25%), where the candidate invasion suppressors genes E-cadherin and α-catenin are located, respectively (see below). Furthermore, frequent loss of chromosome 13 has been reported and this may involve loss of the retinoblastoma (RB) gene or a gene associated with inherited breast cancer (BRCA2). Also frequent loss of chromosome 10 q has been described and recently, a putative target gene on this chromosomal arm was identified, namely PTEN/MMAC-1, a protein tyrosine phosphatase gene which is involved in Cowden disease [36, 38, 56]. Clearly, the relevant suppressor genes in the affected chromosomal regions are candidates to be used as molecular markers of prostate cancer development and progression.

2.3 Oncogene and tumor suppressor gene involvement in prostate cancer

For prostate cancer, studies have been carried out on typical tumor suppressor genes such as p 53 and retinoblastoma (RB) and the oncogene *ras*, which are frequently affected in other solid tumors. Several studies to determine a potentially important role for activated *ras* genes in prostate cancer showed low frequencies of gene mutations in American patients suggesting that *ras* activation is not a common event in either the initiation or the progression of prostate cancer [12, 23]. In contrast to this, two reports that describe similar studies on prostate cancer tissue from Japanese men suggest that *ras* gene mutations do occur at significant frequencies in this population both in latent carcinoma [31] and clinically manifest disease [17, 32]. This raises the question whether significant differences exist in the genetic events associated with prostate cancer in American vs. Japanese men.

Numerous studies have examined the Rb and p 53 tumor suppressor genes in prostate cancer. The frequency of p 53 mutations in specimens of primary prostate cancer is low in comparison to many other common cancers. However, several recent studies indicate that bone metastatic deposits of prostate cancer frequently exhibit mutant p 53 [14, 18, 55, 62, 73]. Although further studies are necessary to directly address this question, these results suggest that p 53 mutations are late events in the progression of prostate cancer. Initially it was suggested that Rb gene inactivation was important in a subset of prostate adenocarcinomas. However, here as well more studies are required to establish the usefulness of Rb as a marker for prostate cancer [62, 66].

2.4 Microsatellite instability in prostate cancer

The genes involved in DNA repair are a new category of genes that are inclueed in the group of cancer associated genes. Recently, the frequent occurance of mutations in DNA repair enzyme genes, e. g., hMSH2, in the germ line of patients with hereditary non-polyposis colon cancer was described [33] and tumors from such individuals are characterized by literally thousands of somatic mutations, in the form of changes in the repeat length of simple sequence repeats, also known as microsatellite instability [28, 33, 63]. That fidelity of DNA replication and repair can play such a prominent role in human carcinogenesis is not surprising but has been somewhat overshadowed by the search for oncogenes and tumor suppressor genes, as these are considered to be critical growth regulatory targets of mutational events. For prostate cancer, the frequency of microsatellite instability seems to be low, however, certain prostate cancers do manifest the replication error (RER) phenotype [17, 74]. Further analysis is required to determine the role of microsatellite instability in prostate cancer.

2.5 DNA methylation

In addition to deletions, mutations, and chromosomal translocations, alterations in DNA methylation can affect gene expression. Multiple studies suggest that altered methylation patterns are associated with the development of both benign and malignant tumors and may be associated with aging in general [3, 4, 47]. Hypermethylation of CpG islands (areas of the genome rich in the sequence CpG, associated with the 5′ regulatory regions of genes) has been correlated with gene inactivation. In prostate cancer, the methylation status of the promoter of the glutathione S-transferase (GST) pi gene, located on chromosome 11 q13 and encoding an enzyme that is a key part of an important cellular pathway to prevent damage from a wide range of carcinogens, was studied. Lee and coworkers [35] found that hypermethylation occurred in 20 of 20 human prostatic tumors, but not in normal prostatic tissue or in benign hyperplasia of the prostate (BPH). A striking decrease of GSTP1 expression was found to accompany human prostate carcinogenesis. Further analyses of 91 prostatic carcinomas by staining with an anti-GSTP1 antibody revealed that no GSTP1 expression could be detected in 88 of these cancer specimen, suggesting that (hyper)methylation of GSTP1 regulatory sequences leading to decreased GSTP1 expression may be one of the most common genetic alterations yet described for prostate cancer. Currently, studies are being performed to design a GSTP1 CG island methylation assay that may serve as a new molecular diagnostic tool for prostate cancer detection [34].

2.6 Differential gene expression

Regardless of the mechanisms by which it occurs (allelic gain or loss, translocations or hypermethylation), genetic alterations often result in changes in mRNA expression. This offers another approach to identify genes involved in prostate cancer development, namely by identifying genes that are differentially expressed at the mRNA level using differential or substraction hybridization or differential display analysis:

one can screen for genes which are up-regulated (e. g., genes that are overexpressed as a result of, for instance, gene amplification or oncogene activation) as well as for genes which are down-regulated (e. g., genes that are no longer expressed, due to allelic loss, or transcriptional down-regulation/inactivation). The reagents that become available, i. e., cDNA clones, are easy to characterize by nucleotide sequence analysis and can be tested for usefulness in the diagnosis by Northern blot analysis or RNA *in situ* Hybridization (RISH). Differential and subtraction hybridization have been successfully applied to identify genes differentially expressed when comparing normal to malignant tissue. For prostate cancer, differential hybridization was used to study general changes in gene expression. In our laboratory, we identified a number of cDNAs overexpressed in metastasizing rat prostatic cancers [10, 51]. One of those clones was shown to be identical to High Mobility Group protein I(Y) [10]. HMG-I(Y) is a small non-histone nuclear protein implicated in transcription and replication processes. Its overexpression in dedifferentiated, fast-proliferating cells was reported earlier. The value of HMG-I(Y) as progression marker was further investigated by means of RNA *in situ* hybridization which revealed that HMG-I(Y) levels were indicative for a worse prognosis, however, additional value over the well-established grading methods was not evident [59, 60].

Differential display analysis is nowadays a widely-applied and powerful tool to identify and clone differentially expressed genes [37]. Differential display analysis has some major advantages over traditional differential or subtraction hybridization, in that (1) it allows simultaneous comparison of multiple RNA samples; (2) both up- and down-regulated expression of genes can be studied in one experiment; (3) it is less time-consuming; and (4) only small amounts of RNA are required for this technique, allowing the use of the sometimes rare and/or small human specimens. The number of newly identified genes that are up- or down-regulated in prostate cancer development is rapidly increasing. However, their value as markers for prostate cancer and as putative targets/tools for treatment of prostate cancer patients remains to be established. Recently, we identified a cDNA, DD3, which is strongly overexpressed in prostatic carcinomas in comparison to normal and benign hyperplastic tissue of the prostate. Furthermore, it appears that the expression of DD3 is restricted to the prostate, since even when using a sensitive RT-PCR analysis, no expression of DD3 can be detected in 18 other normal human tissues studied. These data suggest that DD3 is the most prostate-cancer specific gene described today [8].

3. Molecular prognostic factors

Despite the tremendous increase in our knowledge of the molecular basis of prostate cancer, relatively few reports are available on the value of these findings for molecular diagnostics.

3.1 E-cadherin/alpha-catenin immunohistochemistry

One of the best-studied candidates as a molecular prognostic factor for prostate cancer is E-cadherin. This molecule mediates Ca^{2+}-dependent cell-cell interactions and has been shown to play a critical role in embryonic development and morphogenesis [58]. In later stages, E-cadherin is specifically expressed in epithelial tissues. The cadherin family of proteins has now been determined to be quite large, with expression found in numerous different cell types. The functional importance of E-cadherin in preserving epithelial tissue integrity was demonstrated by using antibodies against E-cadherin, leading to dissociation of epithelial cell layers in cell culture and, more significantly, to an increased invasive potential of cells. Increased invasive behaviour can also be induced by introducing E-cadherin antisense RNA into cells. Conversely, introduction of E-cadherin cDNA, by means of transfection into invasive cells, resulted in differentiation of carcinomas. These studies provide evidence that E-cadherin can act as an invasion suppressor [21, 72]. Invasiveness may be acquired by down regulation of E-cadherin at the DNA, RNA, or protein level. In recent years, various laboratories have studied the expression of E-cadherin in a great number of different human and animal cell lines and carcinomas and found a reduced expression of E-cadherin correlating with poor differentiation of human tumors, suggesting an important role for E-cadherin in the maintenance of the differentiated phenotype of carcinomas (for a review, see [6]).

The first studies to examine E-cadherin in prostate cancer were carried out in the Dunning rat prostate cancer model system and showed a strong correlation between the lack of E-cadherin and invasive and/or metastatic potential [9]. This correlation was strengthened by the direct observation of the progression of a non-invasive, E-cadherin positive tumor, to a highly metastatic, E-cadherin negative tumor. To extend these observations to human prostate cancer, immunohistochemical analysis of E-cadherin expression was performed in a large series of human prostate cancer specimens [70]. Whereas all benign samples stained with uniform, strong intensity at cell-cell contacts, one-half of the tumors showed reduced or absent E-cadherin protein staining. When compared with Gleason grade, there was a strong association between high grade and aberrant E-cadherin staining. Moreover, within the Gleason score 6, 7, and 8, E-cadherin differentiated two types, i. e., normal staining pattern *versus* aberrant staining (heterogeneous or negative). Importantly, a subsequent study of the potential prognostic significance of E-cadherin staining shows that indeed, aberrant E-cadherin staining is a powerful predictor of poor outcome, both in terms of disease progression and overall survival [69].

Not all types of adhesion deficiencies in invasive carcinomas are caused by decreased E-cadherin expression. Loss of E-cadherin function may also be due to loss of or reduced contact with the cytoskeleton. Interactions between the cytoskeleton and adhesion molecules have been shown to be essential for a variety of cellular functions, including cell-cell and cell-matrix interactions and cell motility. Normally, the highly conserved cytoplasmic domain of E-cadherin associates with three independent proteins, called α, β-, and γ-catenin [46]. Significant changes in the cadherin-catenin complex have been shown to disturb the junctional complex, thereby affecting the greater motility of invasive cells [6]. Preliminary studies, using immunohistochemical analysis on human prostate cancer specimen, showed that α-catenin expression is often down-regulated in high-grade prostate cancer [68]. Furthermore, these studies indicate that α-catenin may provide additional prognostic information for prostate cancer. However, large prospective studies will have to reveal the clinical value of both E-cadherin and α-catenin as prognostic factors.

Finally, the recent observations that β-catenin is part of the Wnt-signalling pathway, interacting with APC and LEF-1, suggest a potential novel mechanism by which a cancer susceptibility gene may act [5, 48, 57].

3.2 Candidate progression markers for prostate cancer

Several new candidate tissue markers for prostate cancer are currently being investigated, including bcl-2, KAI-1, CD44, the androgen receptor, and telomerase.

The bcl-2 gene was identified due to its involvement at the breakpoint of the t (14; 18) interchromosomal translocation occurring in human follicular B-cell lymphomas. This translocation results in the enhanced expression of bcl-2, which can lead to a decrease in the rate of cell death and malignant transformation [27]. Several studies concerning the role of bcl-2 in the development and progression of prostate cancer have shown that bcl-2 is often not expressed in androgen-dependent prostatic cancer cells whereas it was expressed in androgen-independent prostatic cancer cells [43, 16, 2, 42]. These findings suggest that enhanced expression of bcl-2 protein in carcinomas of the prostate is associated with the transition to androgen independence. Also the value of bcl-2 as a target for the treatment of prostate cancer is currently under investigation.

Recently, a metastasis suppressor gene (*KAI-1*) for prostate cancer was found on chromosome 11p11.2 [19]. KAI-1 belongs to a family of leukocyte surface glycoproteins and decreased expression of this gene has been suggested to be involved in malignant transformation of prostate cancers [20, 67], however, more research is necessary to establish the role of KAI-1 in prostate cancer development.

CD44 forms a group of transmembranous glycoproteins that arise due to alternative splicing of a single mRNA. The expression of the different variants (e. g., containing exon v6) has been suggested to correlate with metastasis and poor prognosis in a number of malignancies. Immunohistochemical analysis of the distribution and prognostic value of CD44s, CD44v5, and CD44v6 in radical prostatectomy specimens suggested that loss of CD44s expression in prostate adenocarcinoma predicts a poor prognosis, independent of stage and grade [45]. However, additional studies are required to determine the usefulness of CD44 as a prostate cancer marker.

The role of androgen in normal prostate physiology is unquestioned. However, the role of androgens and *androgen receptors* (AR) in prostate cancer is much less clear. An initial hypothesis that loss of AR gene expression may be important in androgen independent disease was not supported by several studies which showed continued or even elevated AR gene expression in androgen independent tumors [26, 49]. Newmark and collaborators [64] were the first to report a mutated androgen receptor in a clinical specimen of prostate cancer. More recently several groups have shown the occurrence of mutations of the AR in patients with advanced prostatic cancer [61, 64]. It has been suggested that mutations in the androgen receptor in tumors may instead of making the tumor insensitive to androgen, may make tumors supersensitive to androgen, by an as yet undetermined mechanism. It should also be mentioned here that there are reports suggesting that there is a correlation between the number of polymorphic repeats ($(CAG)_n$, and $(GGN)_n$) in the androgen receptor gene not only with the prognosis for the patient [15, 22], but also with the risk to develop prostate cancer [24, 54].

Telomeres form the ends of eukaryotic chromosomes and consist of tandem repeats of the highly conserved G-rich sequence TTAGGG. With every time the chromosome is replicated, the telomeres shorten until the number of telomeres shrinks below a certain threshold and the cell will senesce. Immortalization of human cells is often associated with reactivation of *telomerase*, a ribonucleoprotein enzyme that adds TTAGGG repeats on telomeres and compensates for their shortening and which is normally only active in stem cells. For prostate cancer, several studies have shown that telomerase activity can be detected in prostatic tumors and that telomerase activity may serve as a biomarker for the detection and evaluation of malignant potential of this disease [39, 50, 53].

4. Perspectives in molecular diagnostics

It is clear that by obtaining more information about the different molecular changes occurring during prostate tumor development, we can eventually elucidate the molecular cascade that leads to prostate cancer and this should result in a better assessment of the biological potential of prostate tumors. The exact place of all the new markers in the diagnostic armamentarium can be determined only in well-controlled, prospective clinical evaluations, in which methods are standardized. The multifocality and heterogeneity that are so characteristic of prostate cancer will undoubtedly complicate interpretation, particularly for biopsy material. The intrinsic sampling artifact will often result in biopsies that are poorly representative of the entire malignant process, as mapped in the radical prostatectomy specimens. Ironically, therefore, the value of presurgical molecular diagnostic procedures will be critically dependent on the development of clinical imaging tools that allow the urologist adequately to biopsy the most malignant area in the prostate. Also the application of PCR assays using primers for prostate-cancer specific genes such as PSA, PSMA, and DD3, to detect occult hematogenous micrometastasis in prostate cancer patients (molecular staging [29, 30] requires the analysis of a great number of samples in large well-designed, standardized, prospective studies.

References

[1] Anwar, K., K. Nalakuki, T. Shiraishi et al.: Presence of ras oncogene mutations and human papillomavirus DNA in human prostate carcinomas. Cancer Res. 52 (1992) 5991–5996.

[2] Apakama, I., M. C. Robinson, N. M. Walter, R. G. Charlton et al.: Bcl-2 overexpression combined with p53 protein accumulation correlates with hormone-refractory prostate cancer. Br. J. Cancer *74* (1996) 1258–1262.
[3] Baylin, S. B., M. Makos, J. J. Wu et al.: Abnormal patterns of DNA methylation in human neoplasia: potential consequences for tumor progression. Cancer Cells *3* (1991) 383–390.
[4] Bird, A.: The essentials of DNA methylation. Cell *70* (1992) 5–8.
[5] Behrens, J., J. P. Von Kries, M. Kuhl et al.: Functional interaction of beta-catenin with the transcription factor LEF-1. Nature *382* (1996) 638–642.
[6] Birchmeier, W., J. Behrens: Cadherin expression in carcinomas: role in formation of cell junctions and the prevention of invasiveness. Biochem Biophys Acta *1198* (1994) 11–26.
[7] Bova, G. S., B. S. Carter, M. J. G. Bussemakers et al.: Homozygous deletion and frequent allelic loss of chromosome 8p22 loci in human prostate cancer. Cancer Res. *53* (1993) 3869–3873.
[8] Bussemakers, M. J. G., A. Van Bokhoven, F. M. J. Debruyne et al.: DD3: a new prostate-specific marker, strongly overexpressed in prostatic tumors. J. Urol. *157* (1997) 21.
[9] Bussemakers, M. J. G., R. J. Van Moorselaar, L. A. Giroldi et al.: Decreased expression of E-cadherin in the progression of rat prostatic cancer. Cancer Res. *52* (1992) 2916–2922.
[10] Bussemakers, M. J. G., W. J. M. Van de Ven, F. M. J. Debruyne et al.: Identification of high mobility group protein I (Y) as a potential marker for prostate cancer by differential hybridization analysis. Cancer Res. *51* (1991) 606–611.
[11] Carter, B. S., T. H. Beaty, C. D. Steinberg et al.: Mendelian inheritance of familial prostate cancer. Proc. Natl. Acad. Sci. USA *89* (1992) 3367–3371.
[12] Carter, B. S., J. I. Epstein, W. B. Isaacs: Ras gene mutations in human prostate cancer. Cancer Res. *50* (1990) 6830–6832.
[13] Cher, M. L., D. MacGrogan, R. Bookstein et al.: Comparative genomic hybridization allelic imbalance and fluorescence in situ hybridization on chromosome 8 in prostate cancer. Genes Chromosomes Cancer *11* (1995) 153–162.
[14] Chi, S. G., R. W. De Vere White, F. J. Meyers et al.: P53 in prostate cancer: frequent expressed transition mutations. J. Natl. Cancer Inst. *86* (1994) 926–933.
[15] Choong, C. S., J. A. Kemppainen, Z. X. Zhou et al.: Reduced androgen receptor gene expression with first exon CAG repeat expansion. Mol. Endocrinol. *10* (1996) 1527–1535.
[16] Colombel, M., Symmans, F., S. Gil et al.: Detection of the apoptosis-suppressing oncoprotein bcl-2 in hormone-refractory human prostate cancers. Am. J. Pathol. *143* (1993) 390–400.
[17] Cunningham, J. M., A. L. Shan, M. J. Wick et al.: Allelic imbalance and microsatellite instability in prostatic adenocarcinoma. Cancer Res. *56* (1996) 4475–4482.
[18] Dinjens, W. N., M. M. Van der Weiden, F. H. Schroder et al.: Frequency and characterization of p53 mutations in primary and metastatic human prostate cancer. Int. J. Cancer *56* (1994) 630–633.
[19] Dong, J. T., P. W. Lamb, C. W. Rinker-Schaeffer et al.: KAI 1, a metastasis suppressor gene for prostate cancer on human chromosome 11p11.2. Science *268* (1995) 884–886.
[20] Dong, J. T., H. Suzuki, S. S. Pin et al.: Down-regulation of the KAI 1 metastasis suppressor gene during the progression of human prostatic cancer infrequently involves gene mutation or allelic loss. Cancer Res. *56* (1996) 4387–4390.
[21] Frixen, U. H., J. Behrens, M. Sachs et al.: E-cadherin-mediated cell-cell adhesion prevents invasiveness of human carcinoma cells. J. Cell Biol. *113* (1991) 173–185.
[22] Giovannucci, E. M. J. Stampfer, K. Krithivas et al.: The CAG repeat within the androgen receptor gene and its relationship to prostate cancer. Proc. Natl. Acad. Sci USA *94* (1997) 3320–3323.
[23] Gumerlock, P. H., U. R. Poonamalice, F. J. Meyers et al.: Activated ras alleles in human carcinoma of the prostate are rare. Cancer Res. *51* (1991) 1632–1637.

[24] Hardy, D. O., H. I. Scher, T. Bogenreider et al.: Androgen receptor CAG repeat lengths in prostate cancer: correlation with age of onset. J. Clin. Endocrinol. Metab. *81* (1996) 4400–4405.
[25] He, W. W., P. J. Sciavolino, J. Wing et al.: A novel human prostate specific, androgen regulated homeobox gene (nkx3.1) that maps to 8 p21, a region frequently deleted in prostate cancer. Genomics *43* (1997) 69–77.
[26] Hobisch, A., Z. Culig, C. Radmayr et al.: Distant metastases from prostatic carcinoma express androgen receptor protein. Cancer Res. *55* (1995) 3068–3072.
[27] Hockenbery, D. M.: The bcl-2 oncogene and apoptosis. Semin Immunol. *4* (1992) 413–420.
[28] Honov, Y, M. A. Peinado, S. Makhosyan et al.: Ubiquitous somatic mutations in simple repeated sequences reveal a new mechanism for colonic carcinogenesis. Nature *363* (1993) 558–561.
[29] Israeli, R. S., W. H. Miller, S. L. Su et al.: Sensitive nested reverse transcriptase polymerase chain reaction detection of circulating prostatic tumor cells: comparison of prostate-specific membrane antigen and prostate-specific antigen based assays. Cancer Res. *54* (1994) 6306–6310.
[30] Katz, A. E., C. A. Olsson, A. J. Raffo et al.: Molecular staging of prostate cancer with the use of an enhanced reverse transcriptase-PCR assay. Urology *43* (1994) 765–775.
[31] Konishi, N., T. Enomoto, G. Buzard et al.: K-ras activation and ras p21 expression in latent prostatic carcinoma in Japanese men. Cancer *20* (1992) 327–338.
[32] Konishi, N., Y. Hiasa, T. Tsuzuki, M. Tao et al.: Comparison of ras activation in prostate carcinoma in Japanese and American men. Prostate *30* (1997) 53–57.
[33] Leach, F. S., N. C. Nicolaides, N. Papadopoulos et al.: Mutations of a mut-s homolog in hereditary nonpolyposis colorectal cancer. Cell *75* (1993) 1215–1225.
[34] Lee, W. H., W. B. Isaacs, G. S. Bova et al.: CG island methylation changes near the GSTPI gene in prostatic carcinoma cells detected using the polymerase chain reaction: a new prostate cancer biomarker. Cancer Epidemiology Biomarkers and Prevention *6* (1997) 443–450.
[35] Lee, W. H., R. A. Morton, J. I. Epstein et al.: Cytidine methylation of regulatory sequences near the pi class glutathione-S-transferase gene accompanies human prostate cancer carcinogenesis. Proc. Natl. Acad. Sci USA *91* (1994) 11733–11737.
[36] Li, J., C. Yen, D. Liaw et al.: PTEN, a putative protein tyrosine phosphatase gene mutated in human brain, breast, and prostate cancer. Science *275* (1997) 1943–1947.
[37] Liang, P., A. B. Pardee: Differential display of eukaryotic messenger RNA by means of the polymerase chain reaction. Science *257* (1992) 967–971.
[38] Liaw, D., D. J. Marsh, J. Li et al.: Germline mutations of the PTEN gene in Cowden disease, an inherited breast and thyroid cancer syndrome. Nat. Genet. *16* (1997) 64–67.
[39] Lin, Y., H. Uemura, K. Fujinami et al.: Telomerase activity in primary prostate cancer. J. Urol. *157* (1997) 1161–1165.
[40] MacGrogan, D., A. Levy, D. Bostwick et al.: Loss of chromosome arm 8 p loci in prostate cancer: mapping by quantitative allelic imbalance. Genes Chromosomes Cancer *10* (1994) 151–159.
[41] Macoska, J. A., T. M. Trybus, W. A. Sakr et al.: Fluorescence in situ hybridization analysis of 8 p allelic loss and chromosome 8 instability in human prostate cancer. Cancer Res. *54* (1994) 3824–3830.
[42] McDonnell, T. J., N. M. Navone, P. Troncoso et al.: Expression of bcl-2 oncoprotein and p53 protein accumulation in bone marrow metastases of androgen independent prostate cancer. J. Urol. *157* (1997) 569–574.
[43] McDonnell, T. J., P. Troncoso, S. M. Brisbay et al.: Expression of the protooncogene bcl-2 in the prostate and its association with emergence of androgen-independent prostate cancer. Cancer Res. *52* (1992) 6940–6944.

[44] Newmark, J. R., D. O. Hardy, D. C. Tonb et al.: Androgen receptor gene mutations in human prostate cancer. Proc. Natl. Acad. Sci. USA *89* (1992) 6319–6323.
[45] Noordzij, M. A., G. J. Van Steenbrugge, N. S. Verkaik et al.: The prognostic value of CD44 isoforms in prostate cancer patients treated by radical prostatectomy. Clin. Cancer Res. *3* (1997) 805–815.
[46] Ozawa, M., H. Baribault, R. Kemler: The cytoplasmic domain of the cell adhesion molecule uvomorulin associates with three independent proteins structurally related in different species. EMBO J. *8* (1989) 1711–1717.
[47] Relk, W.: Cancer genetics: imprinting in leukaemia. Nature *359* (1992) 362–363.
[48] Rubinfeld, B., B. Souza, I. Albert et al.: Association of the APC gene product with beta-catenin. Science *262* (1993) 1731–1734.
[49] Ruizeveld-de Winter, J. A., P. J. Janssen, H. M. Sleddens et al.: Androgen receptor status in localized and locally progressive hormone refractory human prostate cancer. Am. J. Pathol. *144* (1994) 735–746.
[50] Scates, D. K., G. H. Muir, S. Venitt et al.: Detection of telomerase activity in human prostate: a diagnostic marker for prostatic cancer. Brit. J. Urol. *80* (1997) 263–268.
[51] Schalken, J. A., S. B. Ebeling, J. T. Isaacs et al.: Down modulation of fibronectin mRNA in metastasizing rat prostatic cancer cells revealed by differential hybridization analysis. Cancer Res. *48* (1988) 2042–2048.
[52] Smith, J. R., D. Freije, J. D. Carpten et al.: A major susceptibility locus for prostate cancer on chromosome I suggested by a genome wide search. Science *274* (1996) 1371–1374.
[53] Sommerfeld, H. J., A. K. Meeker, M. A. Piatyszek et al.: Telomerase activity: a prevalent marker of malignant human prostate tissue. Cancer Res. *56* (1996) 218–222.
[54] Stanford, J. L., J. J. Just, M. Gibbs et al.: Polymorphic repeats in the androgen receptor gene: molecular markers of prostate cancer risk. Cancer Res. *57* (1997) 1194–1198.
[55] Stapleton, A. M. F., T. L. Timme, A. E. Gousse et al.: Primary human prostate cancer cells harboring p*53* mutations are clonally expanded in metastases. Clin. Cancer Res. *3* (1997) 1389–1397.
[56] Steck, P. A., M. A. Pershouse, S. A. Jasser et al.: Identification of a candidate tumour suppressor gene, MMACI, at chromosome 10q23.3 thsat is mutated in multiple advanced cancers. Nat. Genet. *15* (1997) 356–362.
[57] Su, L. K., B. Vogelstein, K. W. Kinzler: Association of the APC tumor suppressor protein with catenins. Science *262* (1993) 1734–1737.
[58] Takeichi, M.: Cadherin cell adhesion receptors as a morphogenetic regulator. Science *251* (1991) 1451–1455.
[59] Tamimi, Y., H. G. Van der Poel, M. M. Denyn et al.: Increased expression of high mobility group protein I(Y) in high-grade prostatic cancer determined by *in situ* hybridization. Cancer Res. *53* (1993) 5512–5516.
[60] Tamimi, Y., H. G. Van der Poel, H. F. M. Karthaus et al.: A retrospective study of high mobility group protein I(Y) as progression marker for prostate cancer determined by in situ hybridization. Br. J. Cancer *74* (1996) 573–578.
[61] Taplin, M. E., G. J. Bubley, T. D. Shuster et al.: Mutation of the androgen-receptor gene in metastatic androgen-independent prostate cancer. N. Engl. J. Med. *332* (1995) 1393–1398.
[62] Theodorescu, D., S. R. Broder, J. C. Boyd et al.: P53, bcl-2 and retinoblastoma proteins as long term prognostic markers in localized carcinoma of the prostate. J. Urol. *158* (1997) 131–137.
[63] Thibodeau, S. N., G. Bren, D. Schaid: Microsatellite instability in cancer of the proximal colon. Science *260* (1993) 816–819.
[64] Tilley, W. D., G. Buchanan, T. E. Hickey et al.: Mutations in the androgen receptor gene are associated with progression of human prostate cancer to androgen independence. Clin. Cancer Res. *2* (1996) 277–285.

[65] Trapman, J., H. F. B. M. Sleddens, M. M. Van der Weiden et al.: Loss of heterozygosity of chromosome 8 microsatellite loci implicates a candidate tumor suppressor gene between the loci D8S87 and D8S133 in human prostate cancer. Cancer Res. *54* (1994) 6061–6064.

[66] Tricoli, J. V., P. H. Gumerlock, J. L. Yao et al.: Alterations of the retinoblastoma gene in human prostate adenocarcinoma. Genes Chromosomes Cancer *15* (1996) 108–114.

[67] Ueda, T., T. Ichikawa, J. Tamaru et al.: Expression of the KAI 1 protein in benign prostatic hyperplasia and prostate cancer. Am. J. Pathol. *149* (1996) 1435–1440.

[68] Umbas, R., W. B. Isaacs P. P., Bringuier et al.: Relation between aberrant alpha-catenin expression and loss of E-cadherin function in prostate cancer. Int. J. Cancer *74* (1997) 374–377.

[69] Umbas, R., W. B. Isaacs, P. P. Bringuier: Decreased E-cadherin expression is associated with poor prognosis in patients with prostate cancer. Cancer Res. *54* (1994) 3929–3933.

[70] Umbas, R., J. A. Schalken, T. W. Aalders et al.: Expression of the cellular adhesion molecular Ecadherin is reduced or absent in high-grade prostate cancer. Cancer Res. *52* (1992) 5104–5109.

[71] Visakorpi, T., A. Kallioniemi, A. C. Syvanen et al.: Genetic changes in primary and recurrent prostate cancer by comparative genomic hybridization. Cancer Res. *55* (1995) 342–347.

[72] Vleminckx, K., L. Jr. Vakaet, M. Mareel et al.: Genetic manipulation of E-cadherin expression by epithelial tumor cells reveals an invasion suppressor role. Cell *66* (1991) 107–119.

[73] Voeller, H. J., L. Y. Sugars, T. Pretlow et al.: P 53 oncogene mutations in human prostate cancer specimen. J. Urol. *151* (1994) 492–495.

[74] Watanabe, M., H. Imai, H. Kato et al.: Microsatellite instability in latent prostate cancers. Int. J. Cancer *69* (1996) 394–397.

[75] Yatani, R., I. Chigusa, K. Akazaki et al.: Geographic pathology of latent prostatic carcinoma. Int. J. Cancer *29* (1982) 611–616.

Dysfunction of the cell-cell adhesion complex in lethal prostate cancer

R. Paul, W. B. Isaacs, G. S. Bova

Cell adhesion molecules

Three major families of cell adhesion molecules can be differentiated, the integrins, the cadherins and the immunoglobulin superfamily. Integrins mediate mainly cell-substrate adhesion, other integrins are involved in cell-cell adhesion and some have both binding capacities. A wide variety of molecules, which are contained in the immunoglobulin superfamily, participate in cell-cell recognition, cellular immunity, neural development, leukocyte trafficking, signal transduction and other functions. Cadherins, which are the major cell-cell adhesion molecules in a variety of tissues including the glandular epithelium of the prostate, play an important role in tissue compaction, cell differentiation and maintenance of tissue, and are thought to be involved in cell migration and cell regulation [11]. Cadherins are transmembraneous glycoproteins, which interact in a calcium-dependent, homotypic fashion. Common to all cadherins is the molecular structure of three protein domains, the extracellular amino terminus, the transmembraneous domain and the intracellular carboxy terminus. The cadherin binding site [10] lies within the extracellular domain and is responsible for selective, calcium dependent cadherin binding in a homophylic fashion. Besides the critical role in developmental processes, this observation is of special interest in the potential explanation of the mechanism of implantation of metastasis in neoplastic diseases. The different tissue distribution of metastasis of various tumors may be related to the interaction of cell-adhesion molecules, which are present in tumor cells and host tissue cells. Additionally alterations in the expression of cell adhesion molecules may be related to loss of adhesive properties of the tumor cells in the primary neoplastic lesions, resulting in tumor invasiveness and circulating tumor cells. The intracytoplasmic C-terminus of the cadherin molecules is the most conserved region of the peptide sequence. This domain contains an important binding site for a family of intracellular proteins, the catenins [7]. Three catenins were identified, α-, β-, and γ-catenin. Catenins bind to the cadherins as shown for the classical cadherins, and mediate the contact between the cell adhesion molecules and the microfilaments of the cytoskeleton. This interaction was demonstrated previously to be necessary for the functioning of the cell-adhesion complex. Truncation of cadherins at the carboxy-terminus results in a substantial loss of the adhesive properties [9], which was shown to be reversed by co-transfection of cadherin-catenin fusion proteins [1], demonstrating that association of catenins to cadherins is a keystep in the function of intact adhesion complexes. In contrast, alterations in catenin molecules can lead to disruption of the cell-cell adhesion, resulting in tumor aggressiveness and invasiveness in neoplastic diseases.

Cadherins and catenins in normal prostate

The prostate is a glandular organ which is composed of two layers of epithelial cells embedded within a fibromuscular stroma. The epithelium consists mainly of the secretory epithelial cell and the basal cells. The secretory cells express abundantly E-cadherin (Fig. 1a) at the areas of cell-cell contact [12].

Prostatic basal cells are rounded cells resting at the basal membrane of the acini. We have shown that basal cells (Fig.1a + b) express E- and P-cadherin [5], whereas the secretory cells express only E-cadherin. Thus P-cadherin is a basal cell marker in the normal prostate, comparable to the cadherin expression in the glandular epithelium of the breast. Both, secretory and basal prostatic cells, exhibit immunoreactivity for α-, β-, and γ-catenin on immunohistochemistry.

In normal prostatic tissue of patients who recieved androgen ablation therapy because of prostate cancer demonstrate atrophic glands which consist of a single epithelial cell layer. These atrophic prostatic cells (Fig. 1c + d) demonstrate E- and P-Cadherin immunoreactivity (own results), suggesting that the basal cell of prostatic epithelium represents the stem cell population in this tissue. No change in the expression of catenins in androgen deprived tissue can be observed (own results). The fibromuscular stroma in the prostate does not show any expression of the known cadherins in normal and androgen deprived tissue.

Cadherins in prostate cancer

Bussemakers et al [3] demonstrated for the first time a strong correlation between lack of E-cadherin expression and the metastatic potential in the Dunning rat model of prostate cancer. E-cadherin positive tumors were non-invasive, whereas E-cadherin negative tumors were highly metastatic. A large series of prostate cancers was examined by immunohistochemistry for E-cadherin expression [5]. About one half of the examined tumors showed reduced or absent staining for E-cadherin compared to uniform strong staining of the glandular epithelium in normal prostate. Further a strong association of E-cadherin expression and Gleason grade of the prostate carcinomas was shown. A subsequent study elucidated the role of E-cadherin as a prognostic marker in prostatic cancer. The correlation between E-cadherin expression and tumor grade was confirmed and further a correlation to tumor stage, metastasis, survival and progression of disease after radical prostatectomy was established [13]. These results demonstrate that E-cadherin expression is a prognostic marker in prostatic carcinoma.

To examine the organization of the E-cadherin mediated cell-cell adhesion the catenins were investigated, which are known to bind to E-cadherin at the cytoplasmic domain by immunoprecipitation experiments in prostate cancer [8]. This experiment revealed, that α-catenin expression is also, like E-cadherin, frequently reduced in prostatic tumors. Thus the cadherin mediated pathway of cell-cell adhesion may become inactivated in a variety of ways.

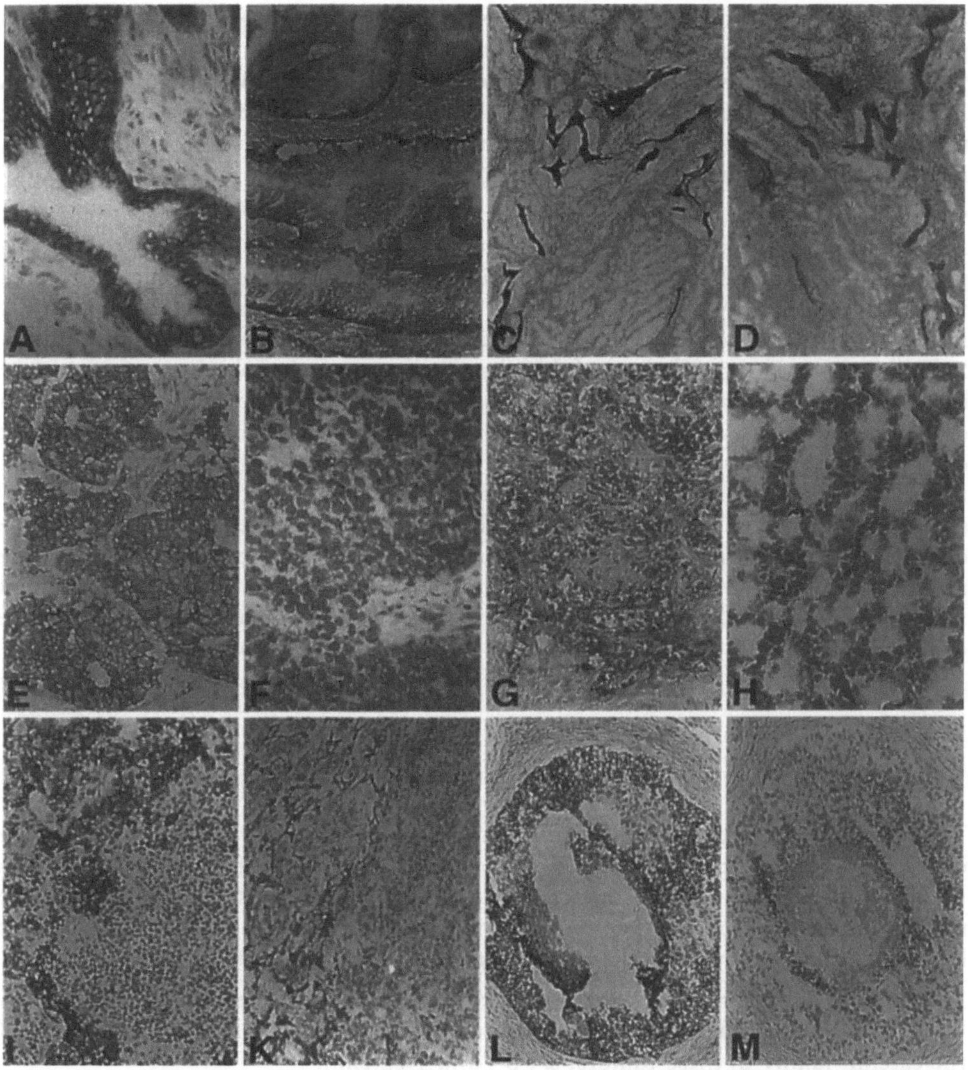

Fig. 1. Immunohistochemistry of normal and neoplastic prostatic tissue for cell adhesion molecules using biotin-streptavidin-method demonstrating in normal prostatic epithelium (A–B).

(A) E-cadherin expression of both basal and secretory cells and (B) restricted P-cadherin expression to the basal cells, compared to androgen deprived atrophic prostatic tissue (C–D) showing positive staining for E-cadherin (C) and P-cadherin (D). Panels E–F demonstrate different immunohistochemical scoring of E-cadherin in different specimens, (E) normal positive staining of an adrenal metastases, (F) heterogeneous expression of a prostatic carcinoma, (G) abnormal staining pattern in a prostate cancer and (H) α-catenin negative lymphnode metastases, demonstrating some positive staining of endothelial cells within the specimen.

(I) demonstrates an E-cadherin positive lymphnode metastases compared to (K) an α-catenin negative liver metastases. (L + N) show a serial section of a lymphnode metastases demonstrating clonal upregulation of N-cadherin in E-cadherin negative areas.

P-cadherin, as mentioned above, is expressed only in the basal cells of the normal prostatic epithelium. As some prostatic cancer cell lines express P-cadherin, like DU-145 and PC-3, the role of P-cadherin in primary prostatic cancer was investigated. None of the examined tumors demonstrated P-cadherin expression [5], suggesting that loss of P-cadherin expression, like other markers of prostatic basal cells [6, 14], is an early event in prostatic cancer.

Other cadherins may be involved in the process of tumor progression and metastasis in prostate cancer. Loss of cadherin expression has been shown to play a role in this process for E-cadherin, but novel expression of cadherins, which are not expressed in normal tissue, may also be a significant step in the carcinogenesis. Cadherin-6 was found in an attempt to characterize cadherin expression in a neonatal prostate epithelial cell line [4]. The significance of this observation is not clear, but one can speculate that expression of novel cadherins may enable the tumor cell to survive and grow in an environment outside the epithelial compartment of the normal prostate, and that emergent novel cadherin expression may play a role in determination the site of formation of metastasis. Evidence of this hypothesis has been demonstrated by Bussemakers et al [2]; they identified OB-cadherin transcripts, a cell adhesion molecule expressed in osteoblasts, in prostatic cancer cell lines. They conclude that this de novo expression of OB-cadherin in prostate cancer could mediate homotypic cell-cell adhesion to osteoblasts in the bone marrow, which could explain the high incidence of bone metastasis in prostate cancer.

Cadherins and Catenins in lethal prostate cancer

To investigate the expression pattern of cadherins and catenins in advanced prostate cancer and prostate cancer metastases we looked at a total of 16 autopsy patients who died of metastasized prostate cancer. All cases were carefully selected and at the time of autopsy, which was kept as short as possible (mean 4h) tumor material was harvested and snap frozen in liquid nitrogen and stored at −70 °C.

Immunohistochemistry was carried out for the classical cadherins E-, N-, and P-Cadherin, as well as cadherin-6. Furthermore we looked at the expression of proteins of the catenin-family in these tumors, α-, β-, and γ-catenin.

Scoring of these tumor specimens were performed by a pathologist and a urologist independently. The grade of positive expression of the cell adhesion molecule was scored positive if more than 90% of all cancer cells (Fig. 1e), heterogeneous if 10 to 90% (Fig. 1f), and negative if less than 10% of all cancer cells showed positive staining (Fig. 1h). Besides the grade of positivity we also looked for abnormal staining pattern (Fig. 1g), if the staining was not related to the cell-cell border, therefore functioning cell adhesion molecules could not be expected.

In this study we looked at a total of 86 specimens, 8 primary prostate tumors, 2 local recurrences after radical prostatectomy and 76 metastases. Of these 76 metastases we included 23 bone metastases, 32 lymph node metastases, and 21 organ metastases mainly liver, adrenal, lung and subdural metastases.

As expected from studies mentioned above loss of functional E-cadherin expression in metastasized prostate cancer is fairly common and could be observed in 59% of all investigated tumor specimen. Normal expression of E-cadherin could only be observed in 41% of metastases (Fig. 1i) and dysfunction of one or more members of the catenin family was found in all of these specimen.

As observed in primary prostate cancer (see above) loss of P-cadherin is common to nearly all metastases (97%), emphasizing that loss of the basal cell compartment in prostate cancer is an early event in tumor progression.

Clonal positivity of N-cadherin in 20% and Cadherin-6 in 4% could be identified in areas where no E-cadherin expression could be detected by immunohistochemistry (Fig. 1l + m). Neoexpression of N-Cadherin and cadherin-6 however seems to play no major role in prostate cancer progression, and expression of these cadherin molecules could not be correlated to a specific site of metastases nor to more favourable or agressive phenotype of this disease.

α-catenin expression was completely lost in 25% (Fig. 1k), dysfunctional either due to abnormal expression or heterogeneous staining was observed in all other tumor specimens. Therefore dysfunction of α-catenin seems to play a crucial role in prostate cancer progression.

β-and γ-catenin also demonstrated loss of expression or abnormal expression in a high percentage of metastases, compareable to those reported for E-cadherin. Interestingly, dysfunctional β-catenin was more often observed if E-cadherin expression was altered, on the other hand most of E-cadherin normal positive specimen demonstrated normal positive immunoreactivity for β-catenin.

In this study we could not observe a specific pattern of cadherin and catenin expression to a specific site of metastases, on the other hand the pattern of cadherin and catenin dysfunction in the same patient showed almost the same distribution in all metastases of this patient, suggesting a clonal origin. Therefore alterations of cell–cell adhesion may precede tumor metastases as hypothesized above. Neo-expression of the investigated cadherins, N-cadherin and cadherin-6, do not predict the site of the metastases, but we can not rule out that other cadherins like OB-cadherin may be involved in site specific metastases.

Conclusions

Our study looking at dysfunction and neo-expression of cadherin and catenin molecules, as well as earlier studies focussing on cadherin expression in primary prostatic carcinomas, emphasize the important role of these adhesion molecules in prostate cancer progression. Especially loss of the functional cell-cell-adhesion complex seems to play a crucial role in tumor invasiveness and metastases as described above. Further investigations are necessary to fully elucidate the role of these adhesion molecules in prostate cancer and to study the possibly therapeutic options of these findings.

References

[1] Behrens, J., M. M. Mareel, W. Birchmeier et al.: Dissecting tumor cell invasion: Epithelial cells acquire invasive properties after the loss of uvomorulin-mediated cell-cell adhesion. J. Cell. Biol. *108* (1989) 2435–2447.

[2] Bussemakers, M. J. G., X. Dikhoff, A. van Bokhoven et al.: De novo expression of osteoblast (OB)-cadherin/11 in prostate cancer. J. Urol. *155* (1996) 351A.

[3] Bussemakers, M. J. G., L. A. Giroldi, F. M. F. Debruyne et al.: Decreased expression of E-cadherin in the progression of rat prostatic cancer. Cancer Res. *52* (1992) 2916–2922.

[4] Ewing, C. M., W. B. Isaacs, M. J. Bussemaker et al.: Analysis of a cadherin-6 like molecule found in neonatal and neoplastic but not in the normal adult prostate tissue. J. Urol. *155* (Suppl.) (1996) 694A.

[5] Jarrard, D. F., R. Paul, A. van Bokhoven et al.: P-cadherin is a basal cell specific epithelial marker that is selectively lost in prostate cancer. Submitted for publication.

[6] Kleinerman, D. I., P. Tronsco, S. H. Lin et al.: Consistent expression of an epithelial cell adhesion molecule (C-CAM) during human prostate development and loss of expression in prostate cancer: implication as a tumor suppressor. Cancer Res. *55* (1995) 1215–1220.

[7] Ozawa, M., R. Kemler: Molecular organization of the uvomorulin-catenin complex. J. Cell. Biol. *116* (1992) 989–996.

[8] Morton, R. A., C. M. Ewing, W. B. Isaacs et al.: Reduction of E-cadherin levels and deletion of the alpha-catenin gene in human prostate cancer cells. Cancer Res. *53* (1993) 3585–3590.

[9] Nagafuchi, A., M. Takeichi: Cell binding function of E-cadherin is regulated by the cytoplasmic domain. EMBO J. *7* (1988) 3679–3684.

[10] Nose, A., K. Tsuji, M. Takeichi: Localization of speicificity determing sites in cadherin cell adhesion molecules. Cell *61* (1990) 147–155.

[11] Takeichi, M.: Cell-cell adhesion molecules controlling animal morphogenesis. Development *102* (1988) 639–655.

[12] Umbas, R., J. A. Schalken, T. W. Aalders et al.: Expression of the cellular adhesion molecule E-cadherin is reduced or absent in high-grade prostate cancer. Cancer Res. *52* (1992) 5104–5109.

[13] Umbas, R., W. B. Isaacs, P. P. Bringuier et al.: Decreased E-cadherin expression is associated with poor prognosis in patients with prostate cancer. Cancer Res. *54* (1994) 3929–3933.

[14] Wojno, K. J., J. I. Epstein: The utility of basal cell-specific anti-cytokeratin antibody (34 beta E12) in the diagnosis of prostate cancer. A review of 228 cases. Am. J. Surg. Pathol. *19* (1995) 251–260.

Heparan sulfate proteoglycans and prostate cancer: a conceptual framework

D. H. J. Schamhart, K.-H. Kurth

1. Introduction

Proteoglycans (PGs) are proteins present in large quantities at several locations: extracellularly, in the pericellular matrix of fibroblasts (extracellular matrix, ECM) and in the basement membrane (BM) of parenchymal cells, at the surface of mesenchymal, epithelial, and neural cells, and intracellulary, within secretory vesicles and the endoplasmatic reticulum. PGs with their covalently linked glycosaminoglycans (GAGs) were initially only regarded as proteins forming the ground substance or filler material between components of the ECM where they serve structural, mechanical and filtration functions. During the last decade, however, developments in proteoglycan biology indicate PGs as important, auxiliary molecules or even key elements in the growth, differentiation, and maintenance of epithelial structure. Evidence suggests a pivotal role of PGs in the processes involved in cell-cell and cell-matrix interactions *and* the modulation of (heparin-binding) growth factor activity and availability. PGs have been hardly addressed in the development as well as maintenance of the differentiated structure and secretory function of the prostate or in prostate pathology, i. e., benign prostatic hyperplasia and prostate cancer (PCa) [16, 28]. Here, the potential role of PGs, in particular heparan sulfate proteoglycans (HSPGs), in (prostate) tumorigenesis will be addressed. Including some new data about the effects of free GAGs on cell proliferation and motility of prostate cancer cells, an attempt intending to provide a conceptual framework for elucidating the potential role of PGs in the progression of prostate cancer is offered.

Owing to the imposed space and citation constraints, readers interested in extensive discussions on proteoglycan synthesis (including GAG biosynthesis and modification), catabolism, structure, biochemistry and regulation are referred to a number of reviews [4, 8–11, 13, 15, 20–22, 24, 26–29, 32].

2. Heparan sulfate proteoglycans (HSPG)

2.1 Characteristics

In contrast to glycoproteins, containing branched oligosaccharides, a PG consists of a polypeptide 'core', to which glycan chains, the glycosaminoglycans (GAGs), are covalently linked. The sulfated GAG chains of PGs are of four types of (long) repeating disaccharide units: heparan sulfate (HS), keratan sulfate (KS), chondroitin sulfate

(CS) and dermatan sulfate (DS). One residue of the disaccharide is always hexosamine, usually with sulfate ester groups attached at the 4 or 6 positions, while the other monosaccharide is a uronic acid residue of either D-glucuronic acid or L-iduronic acid (Fig. 1 A) [10, 13].

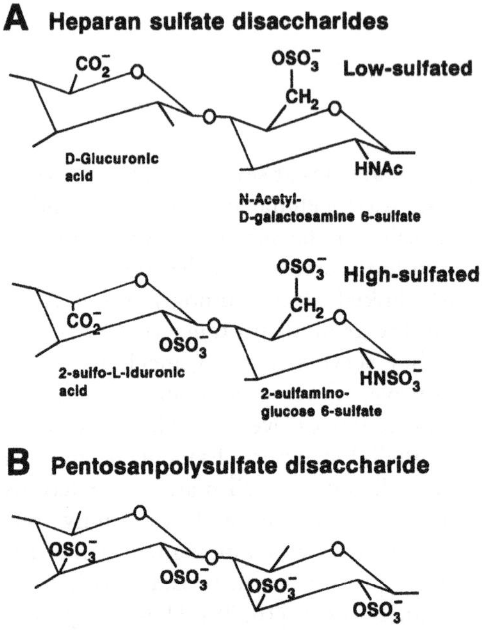

Fig. 1 (A) Structures of disaccharide components as found in heparan sulfate chains of HSPGs. For explanation see text. Note the difference in 'sulfate density' between low- and high-sulfated disaccharides. The low-sulfated-type disaccharides predominate in heparan sulfate. (B) Semi-synthetic pentosanpolysulfate (PPS) derived from beechwood shavings. The structure of PPS, consisting of a highly sulfated ß1→4-D-xylopyranose "backbone", is chemically comparable to GAGs, in particular heparan sulfate. The commercially available preparation (Fibrezym®) contains 6 disaccharide units (Mw ≈ 6,000 D). For comparison, the Mw of a typical PG bound HS chain is approximately 30,000 D.

The GAG chains are posttranslationally added at serine residues of the peptide core through a specialized linkage oligosaccharide. Except in the urinary tract, glycosaminoglycans are not usually found as free chains [10]. Due to the highly complex nature and large diversity, PGs are not very well characterized and there is no uniformly accepted classification, but classification according to the kind of GAG chain(s) is useful, such as HSPGs or DSPGs.

Of the large variety of proteoglycans, HSPGs contain one or more heparan sulfate chain(s), but additional GAGs, such as chondroitin sulfate chains, may be present. The structure of the HS chains consists of disaccharides with a D-glucuronic or L-iduronic acid residue and a glucosamine residue, either as 2-deoxy-2-acetamido- or 2-deoxy-2-sulfamato-glucose, at varying proportions. Furthermore, the HS chains seem to consist of blocks of high-sulfated and blocks of low-sulfated heparan sulfates and this arrangement in blocks is supposed to be of major physiological significance (Fig. 1 A) [10, 32]. Possibly due to their large negative charge density, variation in disaccharide sequence, degree of N- and O-sulphation, acetylation, and C-5 epimerisation of the HS *and* extensive variation in amino acid sequence, size of the core protein and number of attached HS chains, many functions, which comprehend a large range of physiological processes, have been attributed to HSPGs.

2.2 Families of HSPGs

HSPGs are a ubiquitously distributed, highly diverse group of proteoglycans, demonstrated to be present in many mammalian tissues, on cell surfaces and in the ECM and BM. HSPGs can be divided in cell surface-associated HSPGs, consisting of the syndecan and glypican family and an extracellular, BM located PG, perlecan [4, 11, 18, 20].

Syndecans are integral membrane HSPGs and consist of a gene family of four proteins, designed syndecan-1 to -4 or syndecan, fibroglycan, N-syndecan and ryudocan (or amphiglycan), respectively [4, 20]. The core protein of syndecans has a highly conserved transmembrane-spanning region and a short cytoplasmic domain, that communicates with intracellular components both by interaction with the actin cytoskeleton microfilaments and protein kinase C, allowing signal transmission, received from the environment, by the HS chains linked to the extracellular protein domain. The composition of the extracellular core protein differs considerably between the various syndecans, but the HS attachment regions appear to be homogeneous. The various members of the syndecan family appeared to be expressed in distinct cell-, tissue- and development-specific patterns, but no information on prostate tissue is available. Nowadays, most progress on the physiological functions of syndecans has been made with syndecan-1, although knowledge about the other syndecans is rapidly accumulating [2, 4, 6, 12, 20, 25]. Particular regions within the HS chains of syndecan-1 interact with a variety of BM/ECM components, including types I, III, IV and V collagen, fibronectin, thrombospondin, laminin, and tenascin [25]. Along with its basolateral cell surface localization and expression after cell attachment suggest this HSPG as a physiologically relevant matrix receptor. Syndecan-1 has been mentioned in relation to cell-matrix interaction and cell-cell organization. Moreover, syndecans bind growth factors, enzymes and protease inhibitors and have been implicated in the regulation of (heparin binding) growth factor activity (see later).

The *glypican* core protein family is comprised of at least four members, glypican-1 to -4 or glypican-1, cerebroglycan, OCI-5 and K-glypican, respectively [20, 29]. Contrary to syndecans, glypicans are anchored to the perpiheral membrane through glycosylphosphatidylinositol (GPI) linkage). The extracellular domain of glypicans possesses a highly compact tertiary structure and does not have a significant amino acid homology with the syndecans. These PGs are selectively expressed on different cell types. Although the current knowledge about this family of HSPGs is limited, compared to the syndecan family, the available information indicates a role analogous to the syndecans. So, recently obtained data suggest a role of glypican-3 in the modulation of insulin-like growth factor 2 (IGF2) interaction with the IGF2 receptor (IGF2R) [29], comparable to the role of syndecans in facilitating the association of basic fibroblast growth factor (bFGF) with the signal transducing bFGF receptor (see later).

Perlecan is, in addition to the CS/DSPGs decorin and biglycan, the major extracellular PG [18]. This HSPG has been characterized as a large (Mw ≈ 460 kD) protein and multiple splice variants are known. It poses 3 HS chain attachment sites, localized at the N-terminal domain of the core protein and domains, homologous to the LDL-receptor, laminin short arm, N-CAM and laminin A globular end. Perlecan is able to bind a multitude of proteins, ranging from the BM constituents laminin, collagen IV, and fibronectin to growth factors. Its function has been related to the assembly of the BM, involvement in cell attachment/detachment via integrin and non-integrin cell surface receptors (two Leu-Arg-Glu sequences are present in the core protein) and as storage molecule or reservoir of growth factors, like bFGF (see below).

2.3 Functions of HSPGs

2.3.1 HSPGs and growth factor activity

The physical characteristics of HSPGs are consistent with the property to bind heparin-binding growth factors. Indeed, a large variety of growth factors has been currently recognized to bind to HSPGs in addition to heparin. These growth factors comprehend fibroblast growth factors (aFGF, bFGF, K-FGF), hepatocyte growth factor/scatter factor (HGF), vascular endothelial cell growth factor/vascular permeability factor (VEGF/VPF), pleiotrophin (PTN) and others. The binding to HSPGs seems to further several purposes: (i) it may serve in the protection to proteolytic degradation of growth factors, (ii) the association PG-growth factor can provide a matrix and/or cell surface reservoir (depot function) or space/time control of growth factors by (controlled) enzymatic release from the associated PG, and (iii) PG-growth factor complex formation can functionally modulate the activity of the growth factor.

The importance of growth factor-HSPGs interaction is presented in Fig. 2, illustrating the HSPGs-mediated control of basic fibroblast growth factor (bFGF; FGF-2) activity, the prototype example of PG-mediated control of growth factor activity [22, 26, 33]. HSPGs have been identified as necessary molecules to present bFGF to its high affinity, signal-transducing, tyrosine kinase receptor, FGFR1. The HS chains associated with syndecans-1, -2, and -4 as well as glypican-1 bear specific domains for binding of FGF *and* interaction with the high affinity FGF receptor. HSPG-mediated dimerization of bFGF and presentation to FGFR1 appears to be essential for the binding of bFGF to FGFR1, required to elicit appropriate biological responses. Additional regulation of FGFs refers to the ability of the basement membrane HSPG, perlecan to bind FGFs. The significance of the interaction of perlecan with bFGF, which reduces the physiological availability of bFGF, seems to be the provision of a ready available reservoir of bFGF *and* a protection mechanism for bFGF degradation.

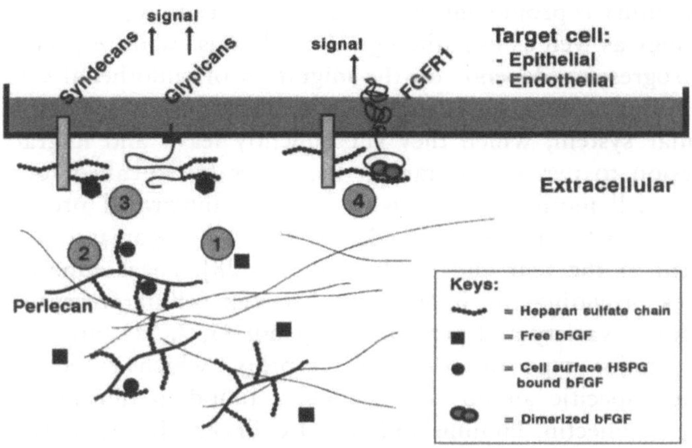

Fig. 2 Schematic representation of the regulation of the growth factor activity and bioavailability of bFGF by heparan sulfate proteoglycans. ① Extracellular bFGF, not bound to perlecan and probably highly sensitive to proteolytic degradation; ② Perlecan-associated bFGF serving as an extracellular reservoir; ③ bFGF bound to syndecan and/or glypican, possibly being a pericellular bFGF reservoir and/or exhibiting signal transmission independent of FGFR1; ④ HSPG facilitated dimerization, activation and presentation of bFGF to FGFR1.

The illustrated, varied regulatory effects of HSPGs may be of equal importance for other growth factors. It seems highly feasible that the bioavailability of other heparin- and non-heparin-binding growth factors may be regulated in a more or less comparable fashion. So, the ECM-associated dermatan sulfate proteoglycan decorin has been demonstrated to bind and neutralize TGF-ß, while the cell membrane-associated PG betaglycan or type III TGF-ß receptor appears to mediate cell proliferation by facilitating the interaction of TGF-ß with its signalling receptors [24]. Furthermore, the very recently described regulatory role of glypican-3 in the IGF2-IGF2 receptor interaction underscores the importance of PG in the regulation of growth factor activity [29].

The importance of the outlined regulatory role of HSPGs (and other PGs) could be of general, basic and clinical significance. It may, for example, at least partly explain the not well-defined role of bFGF in PCa growth or the recently reported discrepancy between the effects of TGF-ß on prostate cancer growth depending on the conditions, *in vitro* or *in vivo* [3]. Furthermore, determination of growth factors expression or concentration by immunohistochemical and/or biochemical techniques in patient material or their effects in in vitro model systems may not reflect their physiological activity in vivo, since their actual physiological activity in vivo could depend on their interaction with cell surface and matrix associated proteoglycans.

2.3.2 HSPGs, cell-ECM adhesion and cell motility

Cell adhesion to constituents of the ECM and BM, participates in a large variety of regulatory events including cell motility, cell proliferation, apoptosis, and specific gene expression [23]. Cell motility is prominent in a wide variety of biology phenomena, both in normal physiology as well as in pathological conditions. With respect to the latter, tumor growth progression depends on the migration of endothelial cells into the tumor mass, while during metastasis tumor cells migrate from the initial tumor mass into the circular system, which they subsequently leave and migrate into a new site. Cell adhesion to pericellular matrix molecules by so-called focal cell adhesion sites mediates cell motility, which is a dynamic, integrated process of formation of focal adhesion sites at the front edge of the cell, organization of stress fibers and detachment at the rear end of the cell [17, 28]. Focal adhesion sites are complex structures providing the interactions between cellular receptors of the integrin and non-integrin variety and pericellular ligands [5, 7, 31]. Integrins are considered to be the major type of adhesion receptors, which recognize, although with low affinities, specific amino acid sequences found in fibronectin, fibrogen, thrombospondin, vitronectin, laminin and type I collagen [1, 23]. Recent evidence, however, indicates cell surface-associated PGs (among others HSPGs) as auxiliary molecules that complement and possibly regulate the more specific integrin-mediated adhesion [4, 7, 9, 11, 13, 22, 31]. It should be noted that Kosir et al. [16], studying the synthesis of HSPGs in a human PCa cell line, PC-3, cultured *in vitro*, demonstrated the presence of various HSPGs in the medium and especially at the cellular adhesion sites. In general, HSPGs are thought to promote cell adhesion, whereas other groups of PGs, the CSPGs and DSPGs, decrease cellular adhesion to extracellular matrix components.

In view of the potential important regulatory role of HSPGs in the motility of tumor cells, the (very limited) number of studies on PGs in normal and malignant prostate tissue addressing this subject [16, 26–28] and our previous data on the effect of added GAGs and PCa cell adhesion and proliferation [28], we initiated a study on the possible relevance of PGs on the *migration* of prostate cancer cells. In a previous study, we showed that the semi-synthetic GAG-like polysaccharide pentosanpolysulfate (PPS) (Fig. 1 B) affects cell adhesion, morphology and proliferation *in vitro* of the human prostatic cell lines LNCaP and DU 145, but not PC-3 cells, when added *during* the process of cell adhesion at several substrata [28]. No inhibitory effects were observed with the natural GAGs, HS, DS, C4S, C6S, and heparin. The results were explained as a competitive interference of PPS with the heparan sulfate chains of the HSPG involved in cell adhesion (and subsequent cell proliferation), due to the high structural similarity of PPS with HS, accompanied by a very high degree of sulfation or charge density (Fig. 1 B). The observed PPS-associated inhibition of LNCaP and DU 145 cell proliferation seems to agree with the reported PPS-induced inhibition of the growth rate of DU 145 in an animal model [34].

Based on the mentioned results of PPS interference with cell adhesion, we studied the effects on cell migration *in vitro* of DU 145 and PC-3 cells in the presence of free GAGs, including natural GAGs, low molecular weight heparin (LMWH) and PPS. As shown in Fig. 3, *in vitro* cell migration of both DU 145 and PC-3 appeared to be highly sensitive to PPS and to a lesser extent to LMWH. The natural GAGs, including heparin, did not interfere with cell migration. These results seem to be in agreement

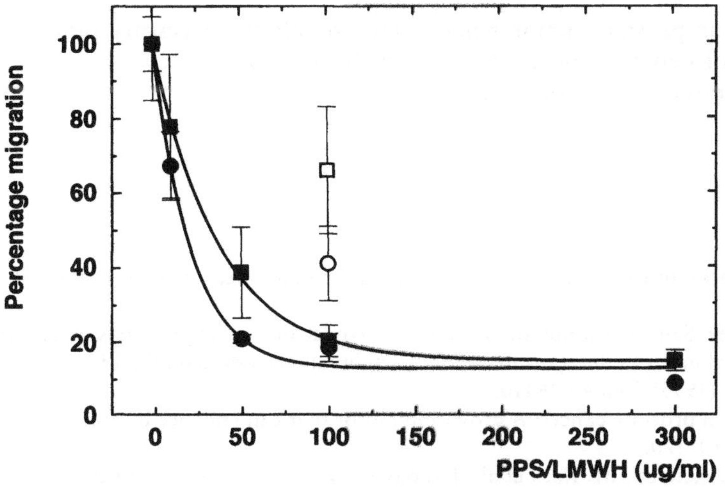

Fig. 3 Effect of pentosan polysulfate (PPS, Fibrezym® and low molecular weight heparin (LMWH, Fraxiparine®) on cell migration of the human prostatic cell lines. Cell migration of DU 145 (*squares*) and PC-3 (*circles*) was quantitated in the presence PPS (10–300 µg/ml) (*closed symbols*) and LMWH (100 µg/ml) (*open symbols*), compared to controls w/o PPS or LMWH, with a Boyden chamber assay using Matrigel® as chemoattractant.

with the observation, obtained by Pienta et al. [19], that PPS disrupts cell-ECM interactions and inhibits motility of a rat prostate adenocarcinoma cell line, MLL, *in vitro*.

3. Perspectives

A number of reports have demonstrated changes in heparan sulfate and HSPGs composition in a variety of human malignancies, although in the majority of cases the exact, *specific* nature still needs to be established [9, 11, 13]. Furthermore, their detailed role in the processes of tumor transformation, progression, and metastasis is far from clear. As outlined in the previous sections, current research in proteoglycan biology is largely focussed on the (modulatory) role of HSPGs (and other PGs) in cell-cell and cell-matrix interaction, and the regulation of the activity and availability of (heparin-binding) growth factors, affecting cell proliferation and motility. Despite the potential importance in tumorigenesis, proteoglycans and their linked glycosaminoglycans have been hardly addressed in prostate cancer. Speculating that the outlined role of proteoglycans is also valid for the progression of prostate cancer, it seems of value to direct major research efforts to characterizing the expression of proteoglycans in prostate cancer cells and tissue, compared to normal tissue. In this respect, the availability of complete nucleotide sequences of a rapidly accumulating number of proteoglycans (syndecans, perlecan and others) and new, more specific antibodies offer new opportunities. Elucidation of the potential role and significance of HSPGs (and other proteoglycans) as molecules modulating growth factor activity and cell motility during prostate tumorigenesis may provide the necessary rationale for the development of new treatment strategies, such as growth factor- or stromal-epithelial interaction-targeted therapies [28, 30].

References

[1] Albelda, S., C. Buck: Integrins and other cell adhesion molecules. FASEB J. *4* (1990) 2868–2880.
[2] Asundi, V., D. Carey: Self-association of N-syndecan (syndecan-3) core protein is mediated by a novel structural motif in the transmembrane domain and ectodomain flanking region. J. Biol. Chem. *270* (1995) 26404–26410.
[3] Barrack, E.: TGFß in prostate cancer: A growth inhibitor that can enhance tumorigenicity. Prostate *31* (1997) 61–70.
[4] Bernfield, M., R. Kokenyesi, M. Kato et al.: Biology of syndecans: A family of transmembrane heparan sulfate proteoglycans. Annu. Rev. Cell Biol. *8* (1992) 365–393.
[5] Burridge, K., K. Fath, T. Kelly et al.: Focal adhesion: Transmembrane junctions between the extracellular matrix and the cytoskeleton. Ann. Rev. Cell Biol. *4* (1988) 487–525.
[6] Carey, D., K. Bendt, R. Stahl: The cytoplasmic domain of syndecan-1 is required for cytoskeleton association but not detergent insolubility. J. Biol. Chem. *271* (1996) 15253–15260.
[7] Gumbiner, B.: Cell adhesion: The molecular basis of tissue architecture and morphogenesis. Cell *84* (1996) 345–357.
[8] Hassel, J., J. Kimura, V. Hascall: Proteoglycan core families. Annu. Rev. Biochem. *55* (1986) 539–567.
[9] Höök, M., L. Kjellén, S. Johansson et al.: Cell-surface glycosaminoglycans. Ann. Rev. Biochem. *53* (1984) 847–869.
[10] Hurst, R.: Structure, function, and pathology of proteoglycans and glycosaminoglycans in the urinary tract. World J. Urol. *12* (1994) 3–10.
[11] Iozzo, R.: Biology of disease. Proteoglycans: structure, function, and role in neoplasia. Lab. Invest. *53* (1985) 373–396.
[12] Itano, N., K. Oguri, Y. Nagayasu: Phosphorylation of a membrane-intercalated proteoglycan, syndecan-2, expressed in a stroma-inducing clone from a mouse Lewis lung carcinoma. Biochem. J. *315* (1996) 925–930.
[13] Jackson, R., S. Busch, A. Cardin: Glycosaminoglycans: Molecular properties, protein interactions, and role in physiological processes. Physiol. Rev. *71* (1991) 481–539.
[14] Kjellén, L., U. Lindahl: Proteoglycans: structures and interactions. Annu. Rev. Biochem. *60* (1991) 443–475.
[15] Kjellén, L., Å. Oldberg, M. Höök: Cell-surface heparan sulfate. Mechanisms of proteoglycan-cell association. J. Biol. Chem. *255* (1980) 10407–10413.
[16] Koshir, M., C. Quinn: Sorting of heparan sulfate proteglycan into matrix compartments of prostate adenocarcinoma cells. J. Surg. Res. *58* (1995) 46–52.
[17] Lauffenburger, D., A. Howitz: Cell migration: A physically integrated molecular process. Cell *84* (1996) 359–369.
[18] Murdoch, A., G. Dodge, I. Cohen et al.: Primary structure of the human heparan sulfate proteoglycan from basement membrane (HSPG2/Perlecan). J. Biol. Chem. *267* (1992) 8544–8557.
[19] Pienta, K., B. Murphy, W. Isaacs et al.: Effect of pentosan, a novel cancer chemotherapeutic agent, on prostate cancer cell growth and motility. Prostate *20* (1992) 233–241.

[20] Rosenberg, R., N. Shworak, J. Liu et al.: Heparan sulfate proteoglycans of the cardiovascular system. Specific structures emerge but how is synthesis regulated? J. Clin. Invest. *99* (1997) 2062–2070.
[21] Ruoslahti, E.: Structure and biology of proteoglycans. Annu. Rev. Cell Biol. *4* (1986) 229–255.
[22] Ruoslahti, E.: Proteoglycans in cell regulation. J. Biol. Chem. *264* (1989) 13369–13372.
[23] Ruoslahti, E., J. Reed: Anchorage dependency, integrins and apoptosis. Cell 77 (1994) 477–478.
[24] Ruoslahti, E., Y. Yamaguchi: Proteoglycans as modulators of growth factor activities. Cell *64* (1991) 867–869.
[25] Sanderson, R., J. Turnbull, J. Gallagher et al.: Fine structure of heparan sulfate regulates syndecan-1 function and cell behavior. J. Biol. Chem. *269* (1994) 13100–13106.
[26] Schamhart, D., K.-H. Kurth: Glycosaminoglycans and prostate cancer: a therapeutic future? Perfusion 7 (1994) 1–6.
[27] Schamhart, D., K.-H. Kurth: Proteoglycans and glycosaminoglycans in tumor growth and migration: First experience with tumors of bladder and prostate origin. World J. Urol. *12* (1994) 55–61.
[28] Schamhart, D., K.-H. Kurth: Role of proteoglycans in cell adhesion of prostate cancer cells: from review to experiment. Urol. Res. *25* (*Suppl. 2*) (1979) S 89–S 96.
[29] Weksberg, R., J. Squire, D. Templeton: Glypicans: a growing trend. Nat. Genet. *12* (1996) 225–227.
[30] Wellstein, A., E. Sale, H. Chung et al.: Growth factors as targets in tumor therapy. Int. J. Pharmacognosy *33* (1995) 35–47.
[31] Woods, A., J. Couchman: Focal adhesions and cell-matrix interactions. Collagen Rel. Res. *8* (1988) 155–182.
[32] Yanagishita, M., V. Hascall: Cell surface heparan sulfate proteoglycans. J. Biol. Chem. *267* (1992) 9451–9454.
[33] Yayon, A., M. Klagsbrun, J. Esko et al.: Cell surface, heparin-like molecules are required for binding of basic fibroblast growth factor to its high-affinity receptor. Cell *64* (1991) 841–848.
[34] Zugmaier, G., M. Lippmann, A. Wellstein: Inhibition by pentosan polysulfate (PPS) of heparin-binding growth factors released from tumor cells and blokkage by PPS of tumor growth in animals. J. Natl. Cancer Inst. *84* (1992) 1716–1724.

Androgen receptor mutations in prostate cancer

H. Klocker, Z. Culig, A. Hobisch, A. Hittmair, G. Bartsch, H. Peterziel, A. C. B. Cato

The androgen receptor, the intracellular signal transducer of androgenic stimuli, is a central regulator of prostate growth and function. It is a member of the superfamily of ligand-activated nuclear receptors that includes the steroid hormone receptors, retinoic acid and thyroid receptors, the vitamin D receptor and several orphan receptors of unknown ligand specificity [11, 24, 39]. The gene encoding the AR is located on the long arm of the X-chromosome [2]. Three functional domains harbor the transactivation function at the N-terminus, the DNA-binding activity in the central region and the ligand-binding function at the C-terminus (Fig. 1). In the N-terminal

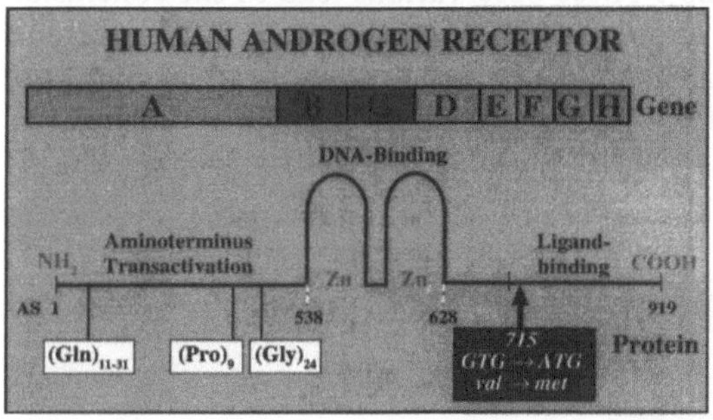

Fig. 1 Structure of the human androgen receptor. The androgen receptor is a member of the superfamily of ligand-activated nuclear transcription factors. The AR gene is located on the long arm of the X-chromosome. The N-terminal transactivation domain is encoded by exon A, the central DNA-binding region built-up of two zink finger elements by exons B and C and the C-terminal ligand-binding domain by exons D–H. The transactivation domain comprises three stretches of homopolymeric amino acid repeats. The point mutation detected in a biopsy derived from an advanced metastatic tumor occured in the ligand-binding domain at amino acid position 715.

region there are three homopolymeric amino acids repeats, a glutamine repeat of variable length, a repeat of nine prolin residues and a polyglycin repeat. An abnormal extension of the glutamine repeat is associated with X-linked spinal and bulbar muscular atrophy (Kennedy Syndrome) [23, 25], and very recent results suggest an association between AR allels with short polyglutamine repeats and an increased risk of prostate cancer [14, 17, 18, 35].

Androgens such as testosterone and dihydrotestosterone bind with high affinity to the ligand-binding domain of the AR, thus inducing a cascade of changes that result in activation of the transcription factor function of the AR. The activation process includes dissociation of heat shock proteins that are associated with the unoccupied receptor, hyperphosphorylation, conformational changes and dimerization of the receptor [22, 34]. The activated AR binds to specific DNA sequences called androgen responsive elements (AREs) located within the promoters of androgen responsive genes [40]. In conjunction with cofactor proteins and other transcription factors, the AR is then able to up- or down-regulate the transcription of genes. A number of genes such as prostate specific antigen (PSA) [31], probasin [30], collagenase I [32], and others have been shown to be regulated by androgens. However, the androgen-regulated genes that perform a crucial role in the control of prostate growth have not been identified yet.

In prostate cancer treatment the AR is the target of endocrine androgen ablation therapy. Blockade of the androgen signal kills prostate cancer cells through induction of so-called "programmed cell death" or apoptosis [8]. Despite a high initial response rate to hormonal androgen ablation therapy its benefits are temporary only because essentially all tumors progress into a therapy-unresponsive state within about two years. In human prostate tumor cell lines and in the rat Dunning tumor prostate cancer model unresponsiveness to therapy is associated with a decrease or absence of androgen receptor expression [6, 9, 28]. These observations lead to the conclusion that a loss of the AR – the target of androgen ablation therapy – is the cause of therapy failure. Therefore it came as a great surprise when immunohistochemical and RT-PCR methods proved expression of the AR in almost all primary and metastatic tumor lesions irrespective of the sensitivity of the tumors to ablation therapy (Fig. 2) [6, 9, 15, 16, 19, 28].

AR EXPRESSION IN PROSTATE CANCER

Number of AR+ cells	SPECIMENS		
	primary tumors n=12	lymph node metast. n=12	distant metast. n=22
-		•	
+	•	••	••••
++	•••••	•	••••••••••
+++	••••••	••••••••	••••••••

- no staining; + <10%, ++ 10-50%, +++ >50% cells stain positive

Fig. 2 Androgen receptor expression in primary and metastatic PCa lesions. Expression of androgen receptor protein was assessed in specimens from primary tumors, lymph node and distant metastases by means of immunohistochemistry and evaluated using a four point scale. Whereas the samples of primary and lymph node metastases derived from untreated patients, the samples of distant metastases all derived from patients with therapy-refractory, progressive disease. All samples except one lymphe node metastasis expressed AR protein. This pattern of expression shows that the AR is present in tumor cells through all steps of tumor development and that there is no correlation between AR expression and responsiveness to androgen ablation therapy.

These findings finally revealed that hormone therapy resistant prostate tumor cells still express a high amount of AR and suggested that mechanisms other than the loss of AR expression are involved in progression of prostate cancer to an androgen-independent state.

Recent investigations in several laboratories provided increasing evidence that crucial events that enable tumor cells to escape therapy are upregulation of AR expression and mutations in the AR gene resulting in aberrant AR function. Genetic studies of primary tumors before and after androgen ablation treatment showed a high degree of AR gene amplifications in one third of recurrent tumors [20, 43]. Gene amplification occurred during therapy indicating that it is the endpoint of a selection process triggered by antiandrogen therapy. Consistent with these observations, it was shown that the androgen-sensitive LNCaP cells cultured for months in steroid-depleted medium up-regulate AR expression [21].

Androgen receptor mutations were identified in a variety of prostate cancer specimens from primary as well as from metastatic tumor sites and in the LNCaP tumor cell line [3, 4, 7, 10, 12, 19, 26, 33, 36, 37, 38, 41, 42]. In our studies we used the reverse transcription polymerase chain reaction (RT-PCR) technique to isolate and sequence AR-cDNA fragments from 10 prostatectomy specimens and from 7 fine-needle aspiration biopsies of advanced tumors in order to screen for structural AR alteration in prostate cancer. In agreement with the immunohistochemical findings AR fragments could be isolated from all radical prostatectomy specimens and aspiration biopsies. DNA sequence analysis of the DNA- and ligand-binding domains revealed no structural changes in 16 of the 17 specimens. In one specimen obtained from a patient with advanced metastatic disease who failed to respond to endocrine and cytotoxic therapy we determined a point mutation in the ligand binding domain of the AR. This G → A mutation changed amino acid 715 of the AR from valine into methionine (codon GTG → ATG) (Fig. 1). To determine whether this mutation was a somatic event that emerged in the course of the disease we analyzed paraffin-embedded archival tissue obtained from this patient by paliative transurethral surgery before the appearance of metastases. These samples contained only wild-type AR sequences indicating that the mutation indeed occurred in association with the progression to a therapy-refractory state.

The mutant androgen receptor AR-met^{715} was recreated in an expression vector and characterized by hormone binding assays and transactivation assays. Its affinity to ligands such as testicular and adrenal androgens, estrogenic and progestagenic steroids and steroidal as well as non-steroidal antiandrogens was unchanged. However, in transactivation assays, this mutant receptor was activated not only by testicular androgens testosterone and dihydrotestosterone but also by the adrenal androgens dehydroepiandrosterone and androstenedione, by progesterone, and by the androgen metabolites androstanediol and androsterone (Fig. 3). The nonsteroidal antiandrogen hydroxyflutamide inhibited the transactivation function of mutant ARmet^{715} at low concentrations in the nanomolar range but activated the mutant receptor at increased concentrations in the micromolar range in the presence as well as in the absence of dihydrotestosterone (Fig. 3). Hormone-independent androgen receptor activation by insulin-like growth factor I (IGF-I) was not changed by the point mutation. In summary this mutant androgen receptor shows a broadened hormonal activation spectrum and it is reasonable to assume that this promiscous ligand specificity is an advantage for tumor cells in an environment deficient of testicular androgens in consequence of androgen ablation therapy (Fig. 4) [4, 5, 27].

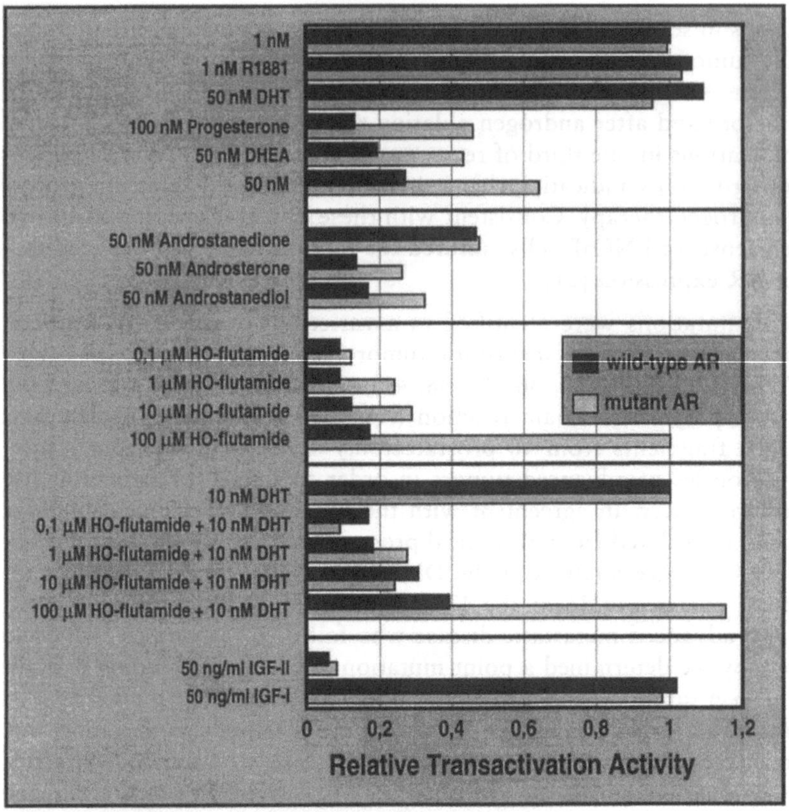

Fig. 3 Transactivation properties of mutant receptor AR-met^{715}. Expression vectors encoding wild-type or mutant ARs and an androgen-inducible reporter plasmid were cotransfected into Du-145 or CV-1 cells by liposome-mediated transfection and the cells were incubated with the hormones, antihormones, or growth factors indicated for 48 h. Transactivation activity was determined by determining reporter gene activity in the cell extracts and calculated relative to the transactivation activity of the wild-type AR activated by saturating concentrations of androgen.

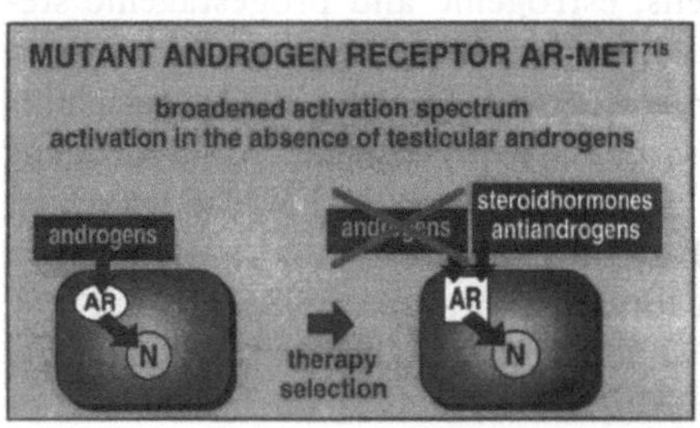

Fig. 4 Summary: mutant androgen receptor AR-met^{715}.
The androgen receptor mutation at amino acid position 715 was identified in a specimen of an advanced therapy-refractory tumor. It was not present in tumor samples obtained before the onset of progression indicating an association between the mutation event and the selection process in the course of androgen ablation therapy. The mutation enables the AR to be activated by a variety of steroid hormones and antiandrogens thus facilitating activation in the absence of testicular androgens.

All together, 26 AR mutations detected in prostate cancer specimens are listed in the human androgen receptor gene mutation database (http://www.mcgill.ca/androgendb/) [13] at the moment. With two exceptions all of these mutations are point mutations changing one amino acid in the AR protein and the majority of these mutations were identified in the ligand-binding domain. The mutation affecting codon 877 that was first discovered in the LNCaP cell line was identified in several tumor specimens, whereas all other mutations were detected only once. Just a few of the mutant receptors have been functionally characterized as of yet. In all of these cases some kind of broadened ligand specificity has been reported suggesting that the mutant ARs in prostate cancer have characteristic similar properties of the kind described for mutant AR-met^{715}. The reported mutated amino acid positions in the ligand-binding domain concentrate into two regions indicating that these parts of the AR determine the ligand specificity of the AR and are hot spots for mutations in prostate cancer (Fig. 5).

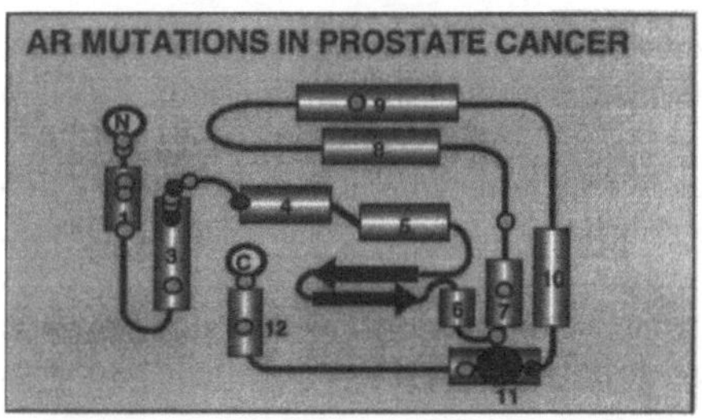

Fig. 5 Amino acid positions in the AR ligand-binding domain mutated in prostate cancer. Several point mutations have been detected in the AR ligand binding domain in prostate cancer specimens. The amino acid positions of these mutations are indicated as circles in a two-dimensional schematic structure of the ligand-binding domain deduced by analogy from the known three-dimensional structure of the ligand-binding domain of a relative of the AR, the retinoic acid receptor-γ. Squares indicate α-helices, arrows β-sheets. Dark circles indicate mutations shown to lead to a promiscuous receptor. The mutation indicated by the bigger circle (amino acid position 877) was identified in several specimens and in the LNCaP tumor cell line. The mutations identified in prostate cancer specimens are concentrated in two regions the first one composed of helices 1, 3, 4, and 12, the second one composed of helices 7 and 11.

With one exception [38], the reports on AR mutations in prostate cancer agree that the frequency of AR mutations in primary prostate tumors is low. In metastatic lesions however Taplin and coworkers reported that mutated receptors are present in almost half of the cases investigated. This finding again underscores the involvement of androgen receptor alterations in prostate cancer progression.

The realization that the androgen receptor is expressed and obviously active during all stages of prostate tumor development has changed our concept of the therapy-refractory cancer. These findings suggest that the androgen signaling cascade is a resonable target for therapy in progressive prostate cancer too, but our current therapeutic means are not sufficient to block androgen stimuli through a hyperactive androgen receptor in these later stages of the disease. It is to be hoped that a better understanding of the structural changes introduced by AR mutations will provide new insights for developing drugs that are able to block also androgen receptors of progressive tumor cells.

References

[1] Bourguet, W., M. Ruff, P. Chambon et al.: Crystal structure of the ligand-binding domain of the human nuclear receptor RXR-alpha. Nature *375* (1995) 377–382.

[2] Brown, C. J., S. J. Goss, D. B. Lubahn et al.: Androgen receptor locus on the human X chromosome: regional localization to Xq 11–12 and description of a DNA polymorphism. Am. J. Hum. Genet. *44* (1989) 264–269.

[3] Castagnaro, M., D. W. Yandell, B. Dworniczak et al.: Androgen receptor gene mutations and p53 gene analysis in advanced prostate cancer. Verh. Dtsch. Ges. Pathol. *77* (1993) 119–123.

[4] Culig, Z. et al.: Mutant androgen receptor detected in an advanced-stage of prostatic carcinoma is activated by adrenal androgens and progesterone. Mol. Endocrinol. *7* (1993) 1541–1550.

[5] Culig, Z. et al.: Activation of two mutant androgen receptors from human prostatic carcinoma by adrenal androgens and metabolic derivatives of testosterone. Cancer Detect. Prev. *20* (1996) 68–75.

[6] Culig, Z., H. Klocker, J. Eberle et al.: DNA sequence of the androgen receptor in prostatic tumor cell lines and tissue specimens assessed by means of the polymerase chain reaction. Prostate *22* (1993) 11–22.

[7] Culig, Z., A. Hobisch, A. Hittmair et al.: Androgen receptor gene mutations in prostate cancer: Implications for disease progression and therapy. Drugs & Aging *1* (1997) 50–58.

[8] Denmeade, S. R., X. S. Lin, J. T. Isaacs: Role of programmed (apoptotic) cell death during the progression and therapy for prostate cancer. Prostate *28* (1996) 251–265.

[9] Diamond, D. A., E. R. Barrack: The relationship of androgen receptor levels to androgen responsiveness in the Dunning R 3327 at prostate tumor sublines. J. Urol. *132* (1984) 821–827.

[10] Elo, J. P., L. Kvist, K. Leinonen et al.: Mutated human androgen receptor gene detected in a prostatic cancer patient is also activated by estradiol. J. Clin. Endocrinol. Metab. *80* (1995) 3494–3500.

[11] Evans, R. M.: The steroid and thyroid hormone receptor superfamily. Science *240* (1988) 889–895.

[12] Gaddipati, J. P., D. G. McLeod, H. B. Heidenberg et al.: Frequent detection of codon 877 mutation in the androgen receptor gene in advanced prostate cancers. Cancer Res. *54* (1994) 2861–2864.

[13] Gottlieb, B., M. Trifiro, R. Lumbroso et al.: The androgen receptor gene mutations database. Nucleic Acids Res. *25* (1997) 158–162.

[14] Hardy, D. O., H. I. Scher, T. Bogenreider et al.: Androgen receptor CAG repeat lengths in prostate cancer: correlation with age of onset. J. Clin. Endocrinol. Metab. *81* (1996) 4400–4405.

[15] Hobisch, A., Z. Culig, C. Radmayr et al.: Distant metastases from prostatic carcinoma express androgen receptor protein. Cancer. Res. *55* (1995) 3068–3072.

[16] Hobisch, A., Z. Culig, C. Radmayr et al.: Androgen receptor status of lymph node metastases from prostatic carcinoma. Prostate *29* (1995) 129–136.

[17] Ingles, S. A., R. K. Ross, M. C. Yu et al.: Association of prostate cancer risk with genetic polymorphisms in vitamin D receptor and androgen receptor. J. Natl. Cancer Inst. *89* (1997) 166–170.
[18] Irvine, R. A., M. C. Yu, R. K. Ross et al.: The CAG and CGC microsatellites of the androgen receptor gene are in linkage disequilibrium in men with prostate cancer. Cancer Res. *55* (1995) 1937–1940.
[19] Kleinerman, D. I. et al.: Expression and structure of the androgen receptor in bone metastases of hormone refractory prostate cancer. J. Urol. *155* (Abstract) (1996) 626 A.
[20] Koivisto, P., E. Hyytinen, C. Palmberg et al.: Analysis of genetic changes underlying local recurrence of prostate carcinoma during androgen deprivation therapy. Am. J. Pathol. *147* (1995) 1608–1614.
[21] Kokontis, J., K. Takakura, N. Hay et al.: Increased androgen receptor activity and altered c-myc expression in prostate cancer cells after long-term androgen deprivation. Cancer Res. *54* (1994) 1566–1573.
[22] Langley, E., Z. X. Zhou, E. M. Wilson: Evidence for an anti-parallel orientation of the ligand-activated human androgen receptor dimer. J. Biol. Chem. *270* (1995) 29983–29990.
[23] LaSpada, A. R., E. M. Wilson, D. B. Lubahn et al.: Androgenreceptor gene mutations in X-linked spinal and bulbar muscular atrophy. Nature 352, 77–79.
[24] Mangelsdorf, D. J., R. M. Evans: The RXR heterodimers and orphan receptors. Cell *83* (1995) 841–850.
[25] Neuschmid-Kaspar, F. et al.: CAG-repeat expansion in androgen receptor in Kennedy disease is not a loss of function mutation. Mol. Cell. Endocrinol. *117* (1996) 149–156.
[26] Newmark, J. R., D. O. Hardy, D. C. Tonb et al.: Androgen receptor gene mutations in human prostate cancer. Proc. Natl. Acad. Sci. USA *89* (1992) 6319–6323.
[27] Peterziel, H., Z. Culig, J. Stober et al.: Mutant androgen receptors in prostatic tumors distinguish between amino-acid-sequence requirements for transactivation and ligand binding. Int. J. Cancer *63* (1995) 544–550.
[28] Quarmby, V. E., W. C. Jr. Beckman, D. B. Cooke et al.: Expression and localization of androgen receptor in the R-3327 Dunning rat prostatic adenocarcinoma. Cancer Res. *50* (1990) 735–739.
[29] Renaud, J. P., N. Rochel, M. Ruff et al.: Crystal structure of the RAR-gamma ligand-binding domain bound to all-trans retinoic acid. Nature *378* (1995) 681–689.
[30] Rennie, P. S. et al.: Characterization of two cis-acting DNA elements involved in the androgen regulation of the probasin gene. Mol. Endocrinol. *7* (1993) 23–36.
[31] Riegman, P. H., R. J. Vlietstra, J. A. van der Korput et al.: The promotor of the prostate-specific antigen gene contains a functional androgen responsive element. Mol. Endocrinol. *5* (1991) 1921–1930.
[32] Schneikert, J., H. Peterziel, P. A. Defossez et al.: Androgen receptor-Ets protein interaction is a novel mechanism for steroid hormone-mediated down modulation of matrix metallo-proteinase expression. J. Biol. Chem. *271* (1996) 23907–23913.
[33] Schoenberg, M. P. et al.: Microsatellite mutation (CAG 24 → 18) in the androgen receptor gene in human prostate cancer. Biochem. Biophys. Res. Commun. *198* (1994) 74–80.
[34] Smith, D. F., D. O. Toft: Steroid receptors and their associated proteins. Mol. Endocrinol. *7* (1993) 4–11.
[35] Stanford, J. L., J. J. Just, M. Gibbs et al.: Polymorphic repeats in the androgen receptor gene: molecular markers of prostate cancer risk. Cancer Res. *57* (1997) 1194–1198.
[36] Suzuki, H., N. Sato, Y. Watabe et al.: Androgen receptor gene mutations in human prostate cancer. J. Steroid Biochem. Mol. Biol. *46* (1993) 759–765.
[37] Taplin, M. E., G. J. Bubley, T. D. Shuster et al.: Mutation of the androgen-receptor gene in metastatic androgen-independent prostate cancer [see comments]. N. Engl. J. Med. *332* (1995) 1393–1398.

[38] Tilley, W. D., G. Buchanan, T. E. Hickey et al.: Mutations in the androgen receptor gene are associated with progression of human prostate cancer to androgen independence. Clin. Cancer Res. *2* (1996) 277–285.

[39] Truss, M., J. Bartsch, G. Chalepakis et al.: Artificial steroid hormone response element generated by Dam-Methylation. Nucleic Acids Res. *20* (1992) 1483–1486.

[40] Truss, M., M. Beato: Steroid hormone receptors: interaction with deoxyribonucleic acid and transcription factors. Endocr. Rev. *14* (1993) 459–479.

[41] Uchida, T., C. Wang, M. Ohori et al.: Infrequent androgen receptor gene mutations in prostate cancer. J. Urol. *155* (abstract) (1996) 624 A.

[42] Veldscholte, J. et al.: A mutation in the ligand binding domain of the androgen receptor of human LNCaP cells affects steroid binding characteristics and response to anti-androgens. Biochem. Biophys. Res. Commun. *173* (1990) 534–540.

[43] Visakorpi, T. et al.: In vivo amplification of the androgen receptor gene and progression of human prostate cancer. Nat. Genetics *4* (1995) 401–406.

[44] Wurtz, J. M., W. Bourguet, J. P. Renaud et al.: A canonical structure for the ligand-binding domain of nuclear receptors [published erratum appears in Nat. Struct. Biol. *3 (2)* (1996) 206]. Nat. Struct. Biol. *3* (1996) 87–94.

Antiandrogen binding to androgen receptors and the consequences

B. J. A. Furr

Introduction

The binding of androgens and antiandrogens to androgen receptors in a range of tissues and the consequences of binding are becoming increasingly complex. Intriguing questions about the reasons for loss of antiandrogen responsiveness by tumours when androgen receptors are still present and the tissue specific nature of gene expression in response to androgen and antiandrogen are slowly becoming partially understood: these questions will be addressed in this paper.

The involvement of a specific intracellular receptor as a mediator of steroid hormone, specifically oestrogen action [13], led to the first simple concepts for androgen action shown in Fig. 1.

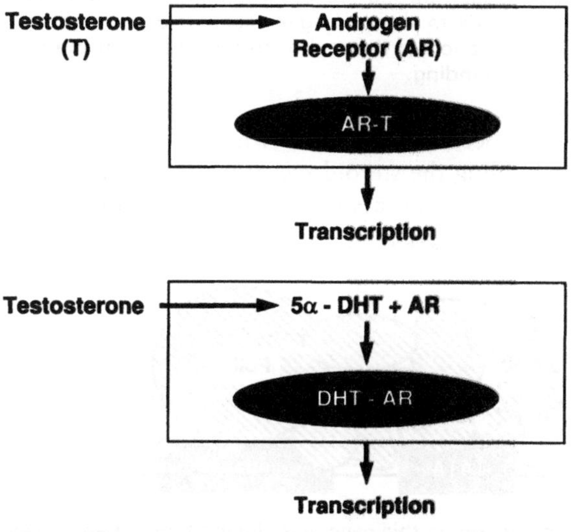

Fig. 1a+b Early concepts of androgen action through receptor binding. Testosterone binds to intracellular androgen receptor that activates transcription by binding to chromatin. Testosterone is converted to 5α-dihydrotestosterone, which is the major ligand binding to the androgen receptor in normal tissue.

Circulating testosterone diffuses into androgen responsive cells of, for example, the prostate and binds to an androgen receptor; the receptor complex then activates gene expression to elicit cell growth and proliferation. It was soon recognised that in normal prostate the major ligand associated with the receptor was a testosterone metabolite. 5α-dihydrotestosterone [1, 4], formed within the prostate from testosterone by the enzyme, 5 α-steroid reductase (Fig. 1). It should be emphasised, however, that testosterone can bind to the androgen receptor directly and effect gene transcription, although it is not as potent as 5 α-dihydrotestosterone. Thus, inhibitors of 5 α-steroid reductase can never effect a complete androgen withdrawal.

More recently, it has been shown that the androgen receptor is normally associated with heat shock proteins and in this state can not bind to hormone response elements on chromatin [19, 41]. Androgen binding causes disassociation of the androgen receptor-heat shock protein complex and exposes a site which allows binding of the androgen receptor to the androgen gene response element (Fig. 2).

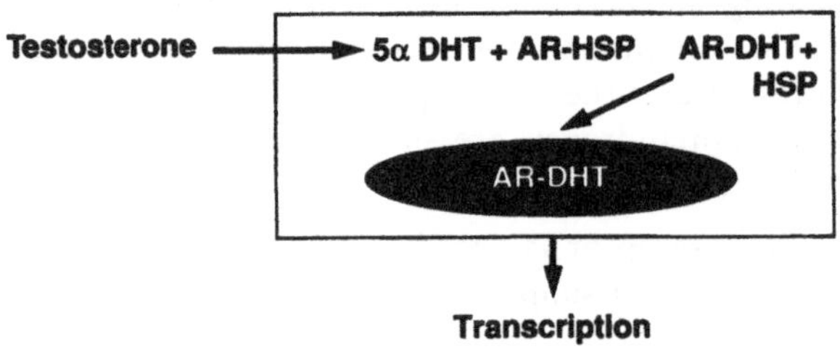

Fig. 2 Evolution of the concept of androgen action through receptor binding. Testosterone is converted to 5α-dihydrotestosterone which binds to the androgen receptor associated with heat shock protein (HSP); this induces disassociation of the androgen receptor from the heat shock protein which allows receptor-chromatin binding.

Work, particularly on oestrogens and glucocorticoids, has shown how naive even this picture has become. Following ligand-binding the steroid receptors disassociate from the heat shock proteins, undergo dimerization and association with a number of co-activating protein transcription factors [22] (Fig. 3).

Fig. 3 Recent concepts of steroid hormone action. In this case, after ligand binding and disassociation from heat shock protein, the steroid receptor (ER) dimerizes and reassociates with a variety of other transcription factors to optimise gene expression. The presence of this galaxy of other factors may convey tissue specificity of response.

It is possible that tissue specificity of response is regulated by the presence or absence of these co-activating factors but this remains to be explored. The simple view that the steroid receptor with its bound hormone and galaxy of co-activators only stimulates gene transcription is also invalid, at least for the glucocorticoid receptor, because transrepression can be demonstrated as well as transactivation. Fig. 4 shows that glucocorticoid receptor can bind to the glucocorticoid response element and induce gene expression.

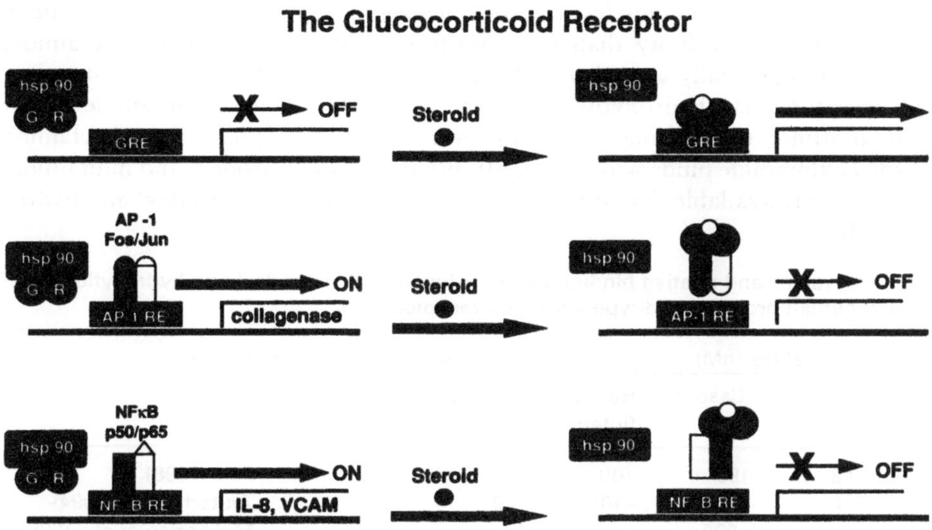

Fig. 4 Steroid receptors can cause both transactivation and transrepression. After disassociation of glucocorticoid receptor (GR) and heat shock protein 90 following ligand binding, the GR can bind to a glucocorticoid response element (GRE) and thereby activate transcription (upper panel). Genes expressed either constitutitively or by stimulation with, for example, tetradecanoyl phorbol acetate (TPA) and regulated by either AP1 (Fos/Jun) or NFκB can be transrepressed when ligand-bound receptor associates with these activating factors and prevents chromatin binding (lower 2 panels).

However, some genes which are expressed either constitutively or in response to various stimuli and are regulated through the AP1 (Fos/Jun) system can be transrepressed by the glucocorticoid receptor which associates with the AP1 complex and prevents its binding to the AP1 recognition elements in chromatin [14]. This may explain why some antiandrogens can repress other growth signals.

The increasingly recognised intricacy of androgen action begins to allow explanations for loss of androgen responses and tissue specific effects but is not the whole story. Improved understanding of the action of androgens and antiandrogens should lead to the design of compounds with subtly improved properties. However, the remainder of this paper will consider only those antiandrogens that are currently available clinically.

Binding of androgens and antiandrogens to wild type androgen receptors

By definition an antiandrogen must bind to the androgen receptor but following binding must not stimulate cell growth and multiplication. It is well established that the 5 clinically available antiandrogens, cyproterone acetate, chlormadinone acetate, flutamide, nilutamide and 'Casodex' all bind to the rat and human prostate androgen receptor but with much lower affinity than the natural ligand 5 α-dihiydrotestosterone. The steroidal compounds (cyproterone acetate, chlormadinone acetate) usually have higher receptor affinity than the non-steroidal drugs (flutamide, nilutamide, 'Casodex'). It is generally agreed that 'Casodex' has higher affinity for prostate androgen receptors than hydroxyflutamide, the active metabolite of flutamide which has similar affinity to nilutamide [11]. However, Luo et al. [21] have recently claimed that hydroxyflutamide binds with higher affinity than both 'Casodex' and nilutamide. Tab. 1 shows the available data on relative binding affinities of 'Casodex' and hydroxyflutamide.

Table 1 IC_{50} values and relative binding affinities for 5α-DHT, 'Casodex' and hydroxyflutamide for rat and human prostate wild-type androgen receptors [11]

Androgen receptor	IC_{50} values (nM)			Relative binding affinities		Author
	5α-DHT	'Casodex'	Hydroxy-flutamide	'Casodex'	Hydroxy-flutamide	
Rat	3.8	190	700	2.0	0.5	Furr (1996)
Rat	~4	~400	~1000	1.0	0.4*	Teutsch et al. (1994)
Rat	–	~200	–	–	–	Mukherjee et al. (1996)
Rat	–	–	–	0.49*	0.15*	Winneker et al. (1989)
Rat	3.6	–	820	–	0.5	Shunsuke et al. (1995)
Rat	17	3595	4565	0.5	0.4	Luo et al. (1996)
Human	–	~100	~300	–	–	Kemppainen & Wilson (1996)
Human	–	–	–	0.3	0.4	Veldscholte et al. (1992)
Human	~6	~300	~700	1.6*	0.7*	Ayub & Levell (1989)
Human	29	2490	2345	1.2	1.2	Luo et al. (1996)

* Values corrected to 5α-DHT = 100%

The majority of studies show that unlike the findings of Luo et al. [21], 'Casodex' has between 2 and 4 times higher affinity for the rat and human receptor than hydroxyflutamide. It is clear from Tab. 1 that the IC_{50} values for 5 α-dihydrotestosterone, hydroxyflutamide and 'Casodex' found by Luo et al. are all considerably higher than values found by other workers. Indeed, the observation that IC_{50} values for 5 α-DHT for human and rat prostate receptors are 29 and 17 nM is surprising since, with ligand excess and under equilibrium conditions, values of a little over 5 nM would be expected because 5 nM [^{3}H] trimethyltrienolone (R 1881) was used as labelled ligand. There may be a simple explanation for the very poor affinity found by Luo et al. [21]: they diluted stock ethanolic solutions of the antiandrogens in buffer that contained no protein. In our experience, dilute solution of drugs frequently bind non-specifically to glass or plastic containers and for that reason we have always used gelatine (protein)-containing assay buffers as diluents.

Luo et al. [21] have also claimed that not only does hydroxyflutamide have higher affinity for the prostate receptor but is a more potent antiandrogen *in vivo* in rats.

Unlike in our studies (Fig. 5), where a good dose-response relation is seen to both flutamide and 'Casodex', inspection of their data shows that, whilst a dose-response is seen to flutamide, there is no difference between the ventral prostate and seminal vesicle weights at doses of 0.4 and 1.2 mg 'Casodex'/Kg, yet both doses suppress organ weights compared with controls.

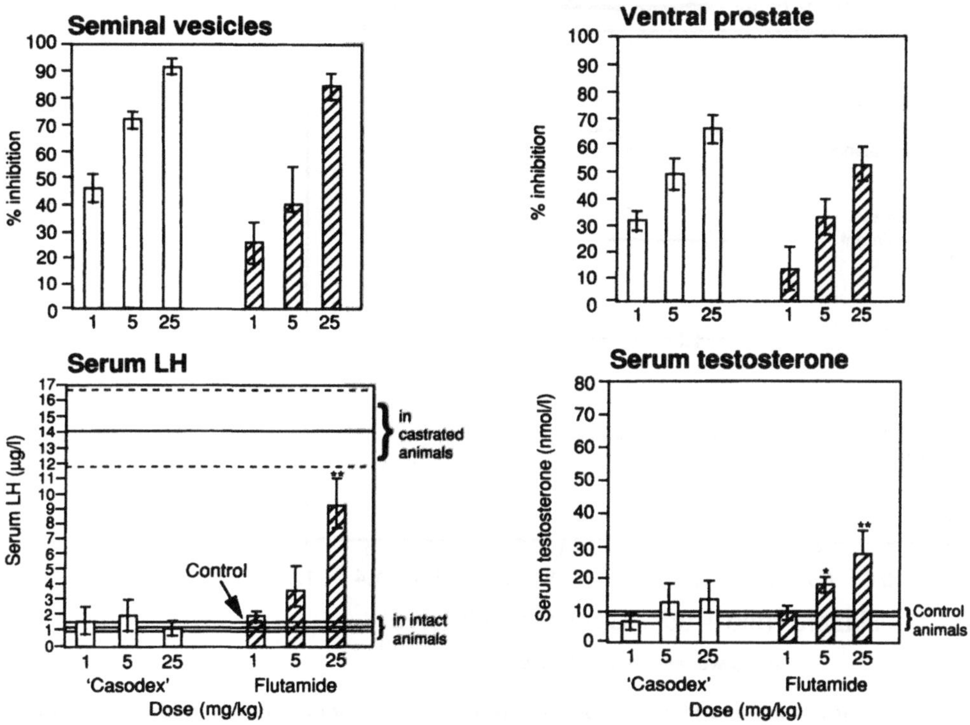

Fig. 5 Dose-response curves for flutamide and 'Casodex'. Both drugs were given orally daily for 14 days. At necropsy on day 15 the ventral prostate and seminal vesicles were weighed and blood was collected for radioimmunoassay of LH and testosterone. The values for organ weights show percentage inhibition compared with surgical castration (100%). The organ weight data show clear dose-responses for both drugs with 'Casodex' being around 5 times as potent as flutamide. The LH and testosterone data show the expected increase with flutamide and the peripheral selectivity (no increase) in the rat of 'Casodex'.

Again there may be a simple methodological explanation for the findings: 'Casodex' has low solubility and was administered by Luo et al. [21] as a suspension. Since particle size can have a profound effect on drug absorption for such insoluble compounds as 'Casodex', we have always used micronised drug to ensure small and consistent particle size so as to optimise bioavailability. There is no information on particle size in the preparation used by Luo et al. [21] and so it is possible that the 'Casodex' used by them had suboptimal bioavailability. There is, however, independent support for the belief that 'Casodex' has higher potency than flutamide in castrated rats given androgen-replacement [36]. Moreover, clinical studies have shown that 'Casodex' is more potent than flutamide and is much better tolerated

[30]. In patients with prostate cancer, LHRH agonist plus 50 mg 'Casodex' is at least as effective as LHRH agonist plus 750 mg flutamide with respect to clinical outcome. Indeed, the hazard ratios of time to treatment failure, time to progression and time to death all favour 'Casodex' although they do not reach statistical significance (Fig. 6).

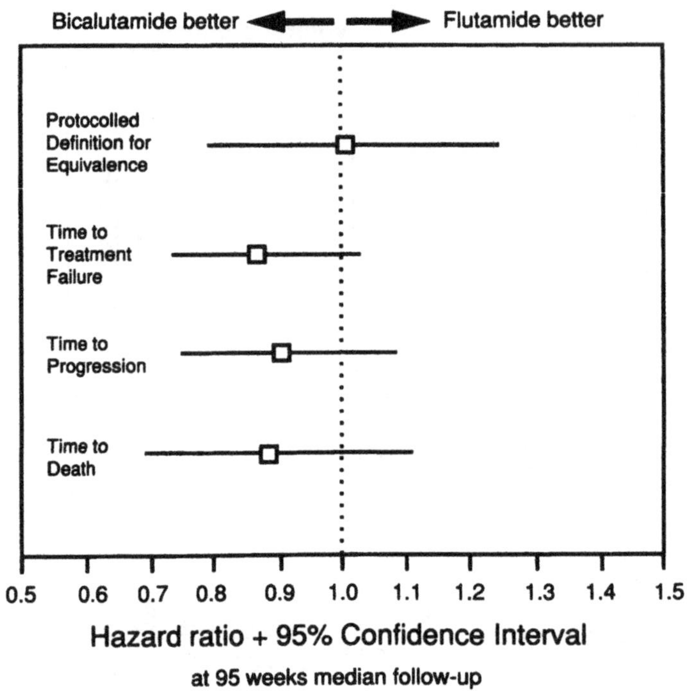

Fig. 6 Hazard ratios and 95% confidence intervals for 'Casodex' (bicalutamide) plus LHRH-A and flutamide plus LHRH-A groups for time to treatment failure, time to progression and time to death.

Antiandrogen insensitivity

There are a number of possible explanations for antiandrogen insensitivity and relapse after an initial response. Apart from changes in rate of metabolism and distribution, three possibilities arise because of changes in androgen receptors. The first is the development of cells which lack the androgen receptor and so are unable to respond to either androgen or antiandrogen. Although this is a common finding with oestrogen receptors in women with breast cancer it has not been as widely observed in prostate cancer. However, a recent report [42] has suggested a significant reduction in androgen receptor expression following complete androgen withdrawal. The second possibility is amplification of androgen receptor expression. Tumours with higher numbers of androgen receptors would give a greater opportunity, even in the presence of low androgen concentrations, for the hormone to bind and effect growth stimulation. Moreover, circulating antiandrogen might not be able to saturate all androgen receptors if present in large numbers. Androgen receptor gene amplification has been demonstrated in about 30% of recurrent tumours but was absent in primary tumours [17]. Such tumours may require higher doses of antiandrogen to effect tumour regression but such tumours also have a more favourable prognosis: the medium survival time was double that for patients with no increase in androgen receptor gene copy number [17].

By far the best documented reason for lack or loss of androgen responses is androgen receptor mutation. Mutations in the androgen receptor have been rarely reported in primary untreated prostate cancer [6, 8, 16, 24, 28, 39]. However, androgen receptor mutations are not uncommon in metastatic prostate cancer and in "androgen insensitive" or relapsing disease [3, 5, 6, 12, 39, 40]. Mutations and/or deletions have been reported in both the DNA and ligand binding domains (see [3] for a comprehensive review). Whilst the functional significance of many of these mutants has not been defined, mutants at codon 706 (Val→Met); 721 (Val→Met); 865 (His-Tyr) and 868 (Thr→Ala; Thr→Ser) have clear functional significance that has a bearing both on poor response to androgen withdrawal and response to oestrogen and progesterone. These mutants also may be responsible for antiandrogen withdrawal responses observed in some patients.

In addition to potent androgens, oestrogens, progestins and adrenal androgens bind to these mutant receptors and stimulate cell growth [5, 6, 27, 37, 40]. Thus, stimulation of the growth of tumours with such receptor mutations would be expected to occur following either castration alone (due to adrenal androgens) or after oestrogen or progestin (cyproterone acetate, chlormadinone acetate) therapy.

These mutated receptors also bind antiandrogens and in the case of flutamide, nilutamide, cyproterone acetate and chlormadinone acetate can stimulate cell growth [41, 5, 37, 33]. However, 'Casodex' still behaves as a pure antiandrogen when bound to the same mutant receptors [5, 7, 37, 41] and fails to cause receptor activation or cell proliferation but blocks growth stimulation by androgens. Mechanistically, it has been shown that hydroxyflutamide induces disassociation of heat shock protein on binding to the 868 (Thr→Ala) mutant found in LNCaP cells which enables the androgen receptor to bind to androgen response elements, whereas 'Casodex' binds to the receptor but fails to effect the disassociation of heat shock protein and is thereby unable to stimulate gene expression [41].

It has been suggested [15] that the increased affinity of hydroxyflutamide for the mutant receptor compared with 'Casodex' causes an increase in stabilisation of the androgen receptor which results in the expression of agonist activity. These workers also report that at high concentrations hydroxyflutamide can also behave as an agonist at the wild type receptor.

The clinical significance of these findings is becoming clear. Flutamide can stimulate tumours in some patients since flutamide withdrawal responses have been observed [9, 10, 20, 29, 32, 35]. There is also preliminary evidence that withdrawal responses to the antiandrogen chlormadinone acetate are associated with the presence of a mutated androgen receptor [2, 38]. The link of activation of mutant androgen receptors and tumour growth is, however, incomplete at this time since 'Casodex', which is an antagonist at those receptors currently investigated, also causes withdrawal responses [25, 34] although possibly with a lower frequency than flutamide [31].

Post receptor binding events

There is growing evidence that protein and peptide growth factors can activate the androgen receptor through a ligand-independent mechanism. Culig et al. [7] have shown that keratinocyte growth factor, insulin-like growth factor 1 and epidermal growth factor can activate androgen receptor in the absence of steroid probably by actions through the protein kinase A and C pathways [18, 23]. This may be a further explanation for apparent androgen insensitivity when tumours continue to grow following medical or surgical castration. However, this is not a true androgen insensitivity since both hydroxyflutamide and 'Casodex' can inhibit growth factor and protein kinase A-stimulated receptor activation [7, 23, 26]. This may explain in part why antiandrogens can be effective following loss of response to castration and may also be part of any benefit of combined androgen withdrawal.

Conclusion

Antiandrogens bind to the androgen receptor with relatively low affinity compared to potent androgens but are still able to block androgen action. Increasing subtleties in action post receptor binding are becoming apparent. 'Casodex' uniquely among currently available antiandrogens, maintains its pure antagonist properties when it binds to clinically expressed androgen receptor mutants; flutamide, nilutamide, cyproterone and chlormadinone acetates act as agonists on binding to some of these mutants and so stimulate tumour growth. The majority of preclinical and clinical studies show that 'Casodex' has greater affinity for the wild type androgen receptor and potency *in vivo* than flutamide. Recent findings show that 'Casodex' can prevent ligand-independent activation of the androgen receptor by protein and peptide growth factors and so prevent tumour growth in a steroid-free environment.

Acknowledgements

I would like to thank Barbara Valcaccia and Jim Woodburn, who were instrumental in the discovery of 'Casodex', and Maurice Needham, who has stimulated my interest in transcription factors and provided some of the illustrations.

References

[1] Anderson, K. M., S. Liao: Selective retention of dihydrotestosterone by prostatic nuclei. Nature (Lond.) *219* (1968) 277–279.

[2] Akakura, K., S. Akimoto, T. Ohki et al.: Antiandrogen withdrawal syndrome in prostate cancer after treatment with steroidal antiandrogen chlormadinone acetate. Urology *45* (1995) 700–705.

[3] Brinkmann, A. O., J.Trapman: Mutant androgen receptors in endocrine-related cancer. Endocrine-Related Cancer *2* (1995) 203–214.

[4] Bruchovsky, N., J. D. Wilson: The intranuclear binding of testosterone and 5 α-androstan-17β-ol-3-one by rat prostate. J. Biol. Chem. *243* (1968) 5953–5960.

[5] Bubley, G. J., S. P. Balk: Treatment of metastatic prostate cancer. Hematol. Oncol. Clin. N. Am. *10* (1996) 713–725.

[6] Culig, Z., A. Hobisch, M. V. Cronauer et al.: Mutant androgen receptor detected in an advanced-stage prostatic carcinoma is activated by adrenal androgens and progesterone. Mol. Endocr. *7* (1993) 1541–1550.

[7] Culig, Z., A. Hobisch, M. V. Cronauer et al.: Androgen receptor activation in prostatic tumour cell lines by insulin-like growth factor-1, keratinocyte growth factor and epidermal growth factor. Cancer Res. *54* (1994) 5474–5478.

[8] Culig, Z., H. Klocker, J. Eberle et al.: DNA sequence of the androgen receptor in prostatic tumour cell lines and tissue specimens assessed by means of the polymerase chain reaction. Prostate *22* (1993) 11–22.

[9] Dupont, A., J. C. Gomez, L. Cusan et al.: Response to flutamide withdrawal in advanced prostate cancer in progression under combination therapy. J. Urol. *150* (1993) 908–913.

[10] Figg, W. D., O. Sartor, M. R. Cooper et al.: Prostate specific antigen decline following discontinuation of flutamide in patients with stage D2 prostate cancer. Am. J. Med. *98* (1995) 412–414.

[11] Furr, B. J. A.: Relative potencies of flutamide and Casodex: Endocrine-related Cancer *4* (1997) 197–202.

[12] Gaddipati, J. P., D. G. McLeod, H. B. Heidenberg et al.: Frequent detection of codon 877 mutation in the androgen receptor gene in advanced prostate cancers. Cancer Res. *54* (1994) 2861–2864.

[13] Jensen, E. V., H. Jacobsen: Basic guides to the mechanisms of oestrogen action. Recent Prog. Horm. Res. *18* (1962) 387–414.

[14] Jonat, C., H. J. Rahmsdorf, K.-K. Park et al.: Antitumour promotion and anti-inflammation: Down modulation of AP-1 (Fos/Jun) activity by glucocorticoid hormones. Cell *62* (1990) 1189–1204.

[15] Kemppainen, J. A., E. M. Wilson: Agonist and antagonist activities of hydroxyflutamide and Casodex relate to androgen receptor stabilization. Urology *48* (1996) 157–162.

[16] Klocker, H., Z. Culig, A. Hubisch et al.: Androgen receptor alterations in prostatic carcinoma. Prostate *25* (1994) 266–273.

[17] Koivisto, P., J. Kononen, C. Palmberg et al.: Androgen receptor gene amplification: A possible molecular mechanism for androgen deprivation therapy failure in prostate cancer. Cancer Res. *57* (1997) 314–319.

[18] Konen, T., J. J. Palvimo, P. J. Kallio et al.: Stimulation of androgen-regulated transactivation by modulation of protein phosphorylation. Endocrinol. *135* (1994) 1359–1366.

[19] Kuil, C. W., C. A. Berrevoets, E. Mulder: Ligand-induced conformational alterations of the androgen receptor analysed by limited trypsinisation. J. Biol. Chem. *270* (1995) 27569–27577.

[20] Liebertz, G., W. K. Kelly, M. Theodoulou et al.: High dose Casodex for prostate cancer (PC): PSA declines in patients (pts) with flutamide withdrawal responses. Proc. Am. Soc. Clin. Oncol. *14* (1995) 232.
[21] Luo, S., C. Martel, G. LeBlanc et al.: Relative potencies of flutamide and 'Casodex': preclinical studies. Endocrine-related cancer *3* (1996) 229–241.
[22] Meier, C. A.: Regulation of gene expression by nuclear hormone receptors. J. Receptor and Signal Trans. Res. *17* (1997) 319–335.
[23] Nazareth, L. V., N. L. Weigel: Activation of the human androgen receptor through a protein kinase A signaling pathway. J. Biol. Chem. *271* (1996) 19900–19907.
[24] Newmark, J. R., D. O. Haldy, D. C. Tonb et al.: Androgen receptor gene mutation in human prostate cancer. Proc. Nat'l. Acad. Sci. USA *173* (1992) 6319–6323.
[25] Nieh, P. T.: Withdrawal phenomenon with the antiandrogen 'Casodex'. J. Urol. *153* (1995) 1070–1073.
[26] Parczyk, K., M. R. Schneider: The future of antihormone therapy: innovations based on an established principle. J. Cancer Res. Clin. Oncol. *122* (1996) 383–396.
[27] Ris-Stalpers, C., M. C. T. Verleun-Mooijman, J. Trapman et al.: Threonine in amino acid position 868 in the human androgen receptor is essential for androgen binding specificity and functional activity. Biochem. Biophys. Res. Comm. *196* (1993) 173–180.
[28] Ruizeveld de Winter, J. A., P. J. A. Janssen, H. F. B. M. Sleddens et al.: Androgen receptor status in localised and locally progressive hormone refractory human prostate cancer. Am. J. Pathol. *144* (1994) 735–746.
[29] Sartor, O., M. Coope, M. Weinberger et al.: Surprising activity of flutamide withdrawal when combined with aminoglutethimide in treatment of "hormone-refractory" prostate cancer. J. Natl. Cancer Inst. *86* (1994) 222–227.
[30] Schellhammer, P., R. Sharifi, N. L. Block et al.: A controlled trial of bicalutamide versus flutamide, each in combination with luteinizing hormone-releasing hormone analogue therapy, in patients with advanced prostate carcinoma. Cancer *78* (1996) 2164–2169.
[31] Schellhammer, P., P. Venner, G. P. Haas et al.: Prostate specific antigen decreases after withdrawal of antiandrogen therapy with bicalutamide or flutamide in patients receiving combined androgen blockade. J. Urol. *57* (1997) 1731–1735.
[32] Scher, H. I., W. K. Kelly: The flutamide withdrawal syndrome: Its impact on clinical trials in hormone-refractory prostate cancer. J. Clin. Oncol. *11* (1993) 1566–1572.
[33] Sharief, Y., E. M. Wilson, S. H. Hall et al.: Androgen receptor gene mutations associated with prostatic carcinoma. Proc. AACR *36* (1995) 1605.
[34] Small, E., P. R. Carroll: Prostate-specific antigen decline after bicalutamide withdrawal: evidence for an antiandrogen withdrawal syndrome. Urology *43* (1994) 408–410.
[35] Small, E. J., S. Srinivas: The antiandrogen withdrawal syndrome. Cancer *76* (1995) 1428–1434.
[36] Snyder, B. W., R. C. Winneker, F. H. Batzold: Endocrine profile of WIN 49596 in the rat: a novel androgen receptor antagonist. J. Steroid Biochem. *33* (1989) 1127–1132.
[37] Stober, J., Z. Culig, A. Gast et al.: Ligand specific modulation of activity of mutant androgen receptors from human prostate cancer: a model for investigating hormone-receptor interactions. AACR special conference: Basic and clinical aspects of prostate cancer, Palm Springs A-23.
[38] Suzuki, H., K. Akakura, A. Komiya et al.: Codon 877 mutation in the androgen receptor gene in advanced prostate cancer: Relation to antiandrogen withdrawal syndrome. Prostate *29* (1996) 153–158.
[39] Suzuki, J., N. Sato, Y. Watabe et al.: Androgen receptor gene mutations in human prostate cancer. J. Steroid Biochem. Mol. Biol. *46* (1993) 759–765.
[40] Taplin, M.-E., G. J. Bubley, T. D. Schuster et al.: Mutation of the androgen receptor gene in metastatic androgen-independent prostate cancer. New Engl. J. Med. *332* (1995) 1393–1398.

[41] Veldschote, J., C. A. Berrevoets, A. O. Brinkmann et al.: Antiandrogens and the mutated androgen receptor of LNCaP cells: differential effects on binding affinity, heat shock protein interaction and transcription activation. Biochemistry *31* (1992) 2393–2399.

[42] De Vere White, R., F. Meyers, S.-G. Chi et al.: Human androgen receptor expression in prostate cancer following androgen ablation. Eur. Urol. *31* (1997) 1–6.

Expression of the MDR1 gene in prostate carcinoma

M. J. Siegsmund, B. Schummer, A. Steidler, M. Greschner, K.-U. Köhrmann, P. Alken

The standard treatment of metastatic prostate cancer is still androgen ablation, which results in a response rate of 70–80% with a duration of about 18 to 24 months. The use of chemotherapy in prostate cancer is limited because of two problems: 1 – The patients that are affected with this cancer are usally old and have reduced organ reserves to withstand a therapy that is combined with multiple side effects [14]. 2 – Prostate carcinomas are highly chemoresistant, which has been attributed to the fact that they grow very slowly, with a high number of cells resting in the G_0 phase of the cell cycle [15] and a growth fraction of only 2% compared, for example, with the teratocarcinoma of the testes which has regularly a growth fraction of about 30 to 40%. However, if this hypothesis were true, undifferentiated, androgen unresponsive tumors, which have a clearly enhanced growth fraction [7], should be more sensitive to chemotherapy than better differentiated tumors. This is, however, not the case, and leads to the question if specific genetically defined resistance mechanisms might be overexpressed in prostate carcinoma.

Since the advent of the first chemotherapeutic agents nitrogen mustard and aminopterin in the nineteen forties, the problem of cytotoxic drug resistance has been recognized, and the then simple techniques were soon tried to define its genetic basis [3]. The first to show nonrandom chromosomal changes associated with drug resistance was June Biedler, who consistantly found homogenously staining regions on chromosome 3 that have later been shown to contain multiple copies of the dihydrofolate reductase gene [2].

With the ongoing development of new cytotoxic drugs, another important aspect of drug resistance became evident. Tumor cells treated with one drug developed resistance not only to this one, but also to a whole variety of other substances which the cells have not encountered before. This kind of cross resistance has later been designated as multidrug resistance or MDR and involves different substances and substance groups, like vinca alcaloids, epipodophyllotoxins, anthracyclines, actinomycin D, and taxol.

In late 1970, Ling and coworkers isolated a 170 kDa protein from membranes of multidrug resistant cells, which they called P-glycoprotein and which turned out to be a membrane-bound pump that effluxes all those drugs, before they can exert their cytotoxic effects [6]. Using highly multidrug resistant cells that were established from the cell line KB-3-1, the responsible gene has been cloned in the laboratory of Michael Gottesman and Ira Pastan and named *MDR*1 [13]. With cDNA probes and monoclonal antibodies to P-glycoprotein, its expression has been studied in a multitude of normal tissues and tumors [4, 5]. Normally, it is expressed in liver, colon, adrenal gland, pancreas, kidney, and capillaries of the testes and brain [4]. It is also expressed in tumors of these organs [4, 5]. Additionally it has been shown, that the expression may increase in some tumors under chemotherapy.

There has also been found some expression in urothelial tissues and transitional cell carcinomas, whereas the expression in prostate tissue and prostate carcinoma has been thought to be low and unfrequent. However, this was based on only a low number of samples that have been examined by relatively unsensitive methods, like northern blotting [4, 5]. Newer results that questioned this opinion came from Theyer et al., and ourselves. Theyer and his collegues found a high doxorubicin resistance in the androgen independent human prostate carcinoma cell lines DU 145 and PC 3, which correlated with a high expression of P-glycoprotein and was reversible by verapamil [10]. We ourselves examined in a pilot study the expression of *MDR*1 mRNA in 11 prostate carcinomas by a sensitive RT-PCR method and found expression in 8 of them [8].

These results prompted us to further investigate MDR in prostate, also under the aspect that drug resistance mediated by P-glycoprotein is reversible by a variety of non-cytotoxic agents, for example calcium antagonists [11, 12].

We examined 37 prostate tissue samples from 29 patients. 25 carcinoma samples were obtained by radical prostatectomy, in 8 cases we also got normal tissue. 4 samples were from progressive androgen independent tumors and were obtained by palliative transurethral resection. The samples were snap frozen in liquid nitrogen and stored in −80 °C until further examination.

All samples were examined histologically and graded according to the UICC criteria.

RNA extraction from all prostatic samples was performed with acidic phenol according to standard protocols, and 1 µg of the total RNA was reverse transcribed with MolV reverse transcriptase in a final volume of 20 µl for one hour at 37 °C. 10 µl of the RT reaction product were used in the following PCR respectively in two different reactions with *MDR*1 and β2 Microglobulin primers with a final volume of 100 µl. The primers had the following sequences:

*MDR*1 5'-AGATCAACTCGTAGGAGTGTC (2421−2442)
5'-GTTTCTGTATGGTACCTGCAA (3194−3215)

β2m ACCCCCACTGAAAAAGATGA (1544−1563)
ATCTTCAAACCTCCATGATG (2253−2262 and 3508−3517)

Amplification with *MDR*1 and β2m-specific primers resulted in a 157 bp and a 120 bp fragment, respectively.

Detection and semiquantitative estimation of the amplified fragments was performed by capillary electrophoresis, since we think that it has some advantages compared to other methods of PCR quantitation: It is sensitive, the semiquantitative estimation of the PCR product can be performed by automated integration of the area under the respective peak, and it is non-radioactive.

The kinetics of PCR follow a sigma shaped curve with increasing cycle number. At low cycle numbers, the product accumulates slowly, and then the reaction reaches a phase where the PCR product doubles with each cycle (exponential phase), to flatten out later. In order to quantify the products it is necessary to use a cycle number where the reactions for both genes proceed in the exponential phase. Therefore calibration reactions with increasing cycle numbers have been performed with β2m and *MDR*1 primers and cDNA from the moderately drug-resistant KB-8-5 cell line. This cell line has been established in Michael Gottesmans laboratory based on the cell line KB-3-1 in colchicin. It displays about threefold higher resistance against doxorubicin and sixfold higher resistance against vincristin compared to the parental KB-3-1 cells [1].

Based on these calibration experiments, the following PCR-reactions were run for 27 cycles with *MDR*1 primers and for 21 cycles with β2m primers. The characteristics of the samples are listed in Table 1. Most of them were T3 tumors and tumor grade G 2.

Table 1 Characteristics of the examined prostate samples

Number of normal tissues	8
Number of tumors	29
Tumor stage	
T1N0M0	0
T2N0M0	5
T3N0M0	20
TxNxM1	4
Grade	
G1	1
G2	18
G3	10
Locally confined	25
Metastasized, androgen independent	4

A clear expression of the *MDR*1 gene could demonstrated in 36 of the 37 samples. In one sample the quotient of *MDR*1/β2m peak areas was a factor 10 lower than in all the other samples. This sample was therefore quoted as negative for *MDR*1 expression. In all examined normal tissues and in 28 of the 29 carcinoma samples we detected *MDR*1 mRNA.

In 11 (38%) of the examined prostate carcinoma samples and 5 normal tissue samples (63%) the expression was higher as in our positive control. In the examined tumor-free areas the *MDR*1 expression was higher, although not significantly, than in the carcinomas. T2 and T3 tumors showed an almost equal expression, whereas in G2 tumors the *MDR*1/β2m quotient was about 25% higher than in G2 tumors and additionally, seven of the ten tumors with an *MDR*1 expression higher than KB-8-5 cells were G2 tumors. The higher *MDR*1 expression in normal tissue compared to tumors and the relation between tumor differentiation and *MDR*1 expression point to a possible functional role of this gene in prostatic tissue. Better differentiated tumors are closer in their phenotype to the normal prostate, whereas dedifferentiation

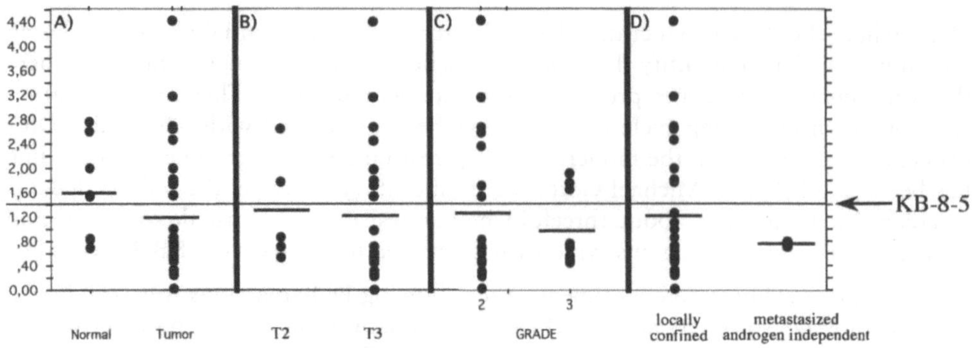

Fig. 1 MDR1/β2m-quotients in prostate samples
Filled circles represent data of each sample, KB-8-5: MDR/β2m quotient of KB-8-5 cells.
A) Comparison of normal and tumor samples
B) Correlation of MDR/β2m-quotient with tumor stage
C) Correlation of MDR/β2m-quotient with tumor grade
D) Comparison of locally confined and metastasized (androgen independent) tumors.

reduces the ability of the cells to express genes that are important for the original function of the gland. A clinically important example is the reduced PSA expression in dedifferentiated prostate carcinomas. We ourselves examined the expression of the IL6 receptor gene in different prostate samples and also found a decreased expression in more dedifferentiated tumors [9].

As stated before, 25 tumors samples were obtained from radical prostatectomy specimens from patients that were not pretreated, and 4 samples were obtained by palliative TUR-P of undifferentiated, androgen independent tumors. By comparison of the means of these two groups we found a clearly higher *MDR*1/β2m quotient in the locally confined tumors than in the metastasized androgen independent and undifferentiated tumors. However, this difference did not reach statistical significance, which most probably is due to the low number of samples in the group of androgen independent tumors and the high variability of values in the locally confined group.

In conclusion, we found that the *MDR*1 gene was expressed frequently in normal prostatic tissue and in prostate carcinomas. This indicates, that it has a physiological role in prostatic function, which is absolutely speculative so far. Additionally, this gene might be involved in the clinically evident drug resistance of prostatic carcinomas which is underscored by the fact that more than one third of the tumors which we examined showed a higher expression than our positive control cells.

Should this be true, it follows that combinations of P-glycoprotein substrates with an inhibitor of P-glycoprotein might be effective in the treatment of some prostate carcinomas.

References

[1] Akiyama, S. I., A. Fojo, M. M. Gottesman et al.: Isolation and genetic characterization of human KB cell lines resistant to multiple drugs. Somatic Cell Mol. Genet. *11* (1985) 117.

[2] Biedler, J.: Drug resistance: genotype versus phenotype-thirty-second G. H. A. Clowes memorial award lecture. Cancer Res. *54* (1994) 666.

[3] Burchenal, J. H., S. F. Johnston, M. N. Kushida et al.: The induction of resistance to 4-amino-N^{10}-methyl-pteroglutamic acid in a strain of transmitted mouse leukemia. Science *3* (1950) 116.

[4] Fojo, A. T., D. G. Poplack, M. M. Gottesman et al.: Expression of a multidrug-resistance gene in human tumors and tissues. Proc. Natl. Acad. Sci. USA *84* (1987) 265.

[5] Goldstein, L. J., H. Galski, A. Fojo et al.: Expression of a multidrug resistance gene in human cancers. J. Natl. Cancer Inst. *81* (1989) 116.

[6] Juliano, R. L., V. Ling: A surface glycoprotein modulating drug permeability in Chinese Hamster ovary cell mutants. Biochim. Biophys. Acta *455* (1976) 152.

[7] Meyers, J. S., G. Sufrin, M. A. Scott: Proliferative activity of benign human prostate, prostatic adenocarcinoma and seminal vesicle evaluated by t hymidine labelling. J. Urol. *128* (1982) 1353.

[8] Siegsmund, M. J., C. Cardarelli, I. Aksentijevich et al.: Ketoconazol effectively reverses multidrug resistance in highly resistant KB cells J. Urol. *151* (1994 a) 485.

[9] Siegsmund, M. J., H. Yamazaki, I. Pastan: Interleukin 6 receptor mRNA in prostate carcinomas and benign prostate hyperplasia. J. Urol. *151* (1994 b) 1396.

[10] Theyer, G., M. Schirmböck, G. Baumgartner et al.: Role of the MDR-1-encodes multiple drug resistance phenotype in prostate cancer cell lines. J. Urol. (1993) 1544.

[11] Tsuruo, T., H. Iida, S. Tsukagoshi et al.: Overcoming of vincristine resistance in P388 leukemia *in vivo* and *in vitro* through enhanced cytotoxicity of vincristine and vinblastine by verapamil. Cancer Res. *41* (1981) 1967.

[12] Tsurno, T., H. Iida, S. Tsukagoshi et al.: Increased accumulation of vincristine and adriamycin in drug-resistant P388 tumor cells following incubation with calcium antagonists and calmodulin inhibitors. Cancer Res. *42* (1982) 4730.

[13] Ueda, K., D. P. Clark, C. J. Chen et al.: The human multidrug resistance (mdr1) gene: cDNA cloning and transcription initiation. J. Biol. Chem. *262* (1987) 505.

[14] Weißbach, L., H. Rübben, W. Jellinghaus: Die Chemotherapie des Prostatakarzinoms. In: R. Nagel: Aktuelle Standortbestimmung der konservativen Therapie des Prostatakarzinoms. de Gruyter, Berlin 1989, 175.

[15] Yagoda, A., D. Petrylak: Cytotoxic chemotherapy for advanced hormone-resistant prostate cancer. Cancer *71* (1993) 1098.

Human prostate cancer progression models and therapeutic intervention

L. W. K. Chung, C. Kao, R. A. Sikes, H. E. Zhau

Abstract

Our laboratory has developed two cellular models of human prostate cancer progression. The LNCaP prostate cancer progression model is based upon the well-known cellular interaction between human prostate or bone stromal cells and LNCaP cells *in vivo*. The marginally tumorigenic LNCaP cells acquired tumorigenic and metastatic potential upon cellular interaction with either prostate or bone fibroblasts. A subline termed C4-2 was observed to grow readily in castrated animals and acquired metastatic potential spreading from the primary tumor site to the lymph node, the seminal vesicles, and the axial skeleton, resulting in an intense osteoblastic reaction. The second model is ARCaP, where prostate cancer cells derived from the ascites fluid of a man with metastatic disease exhibited an *A*ndrogen- and estrogen-*Re*pressed *P*rostate Cancer cell growth and tumor formation in either a hormone-deficient or a castrated environment. However, the growth of either the tumor cells *in vitro* or the tumors *in vivo* was suppressed by both estrogen and androgen. While the tumor cells expressed low levels of androgen receptor and prostate-specific antigen (PSA), they were highly metastatic when inoculated orthotopically. Distant metastases to a number of organs were detected including the liver, lung, kidney, and bone.

We have employed a human prostate cancer progression model as a system to study the efficacy of gene therapy. Results of the study show that whereas universal promoters, such as Cytomegalovirus (CMV) or Rous Sarcoma Virus (RSV) promoter-driven tumor suppressors (e. g., p53, p21 and p16), were effective in inhibiting prostate tumor growth, the advantages of driving the expression of therapeutic toxic genes using a tissue-specific promoter prostate-specific antigen (PSA) and a tumor- but not tissue-specific promoter osteocalcin (OC), are preferred. In the case of the PSA promoter, we can achieve cell-kill in PSA-producing human prostate cancer cells. To circumvent the supporting role of bone stroma for prostate cancer epithelial growth, we have recently developed a novel concept where the expression of therapeutic toxic genes is driven by a tumor- but not a tissue-specific OC promoter. Osteocalcin-thymidine kinase (OC-TK) was found to efficiently eradicate the growth of osteosarcoma, prostate, and brain tumors both *in vitro* and *in vivo*. We observed that androgen-independent human prostate cancer cells lines expressed OC-TK at higher levels than androgen-dependent human prostate cancer cell lines. We have obtained data to suggest that Ad-OC-TK plus a pro-drug acyclovir (ACV) may be used as an effective therapy to treat prostate cancer bone metastasis in models where the growth of androgen-independent PC-3 and C4-2 tumors in the bone has occurred.

Introduction

Prostate cancer is the leading cancer diagnosed in U. S. men and the third leading cause of cancer death in U. S. men [7]. Despite recent efforts to improve treatment of advanced disease using chemotherapy or to prevent human prostate cancer development by the alteration of dietary habits, both incidence and deaths from prostate cancer have risen consistently. In view of the long latency (25–35 years) of prostate cancer before reaching the clinical stage, opportunities are available to develop new therapeutic strategies, both to stage men with prostate cancer and to offer novel treatment therapies in men who harbor metastatic prostate cancer. In our laboratory, we decided first to establish a relevant human prostate cancer model for laboratory investigation. Second, we have developed a molecular staging program to clone novel genes that can be used as molecular markers to differentiate indolent from virulent forms of prostate diseases. Third, we have developed a molecular therapeutic program to devise novel therapeutic approaches to treat human prostate cancers. In this report, we have summarized our recent efforts in the development of animal models to study prostate cancer growth and progression, and the development of an adenoviral therapeutic gene delivery system for the treatment of androgen-independent metastatic prostate cancers in experimental models.

Prostate cancer model development

Out laboratory is responsible for the development of two human prostate cancer xenograft and tissue culture model systems using either LNCaP or newly developed androgen-repressed human prostate cancer cells (ARCaP), grown either *in vitro* or *in vivo*. LNCaP cells, a marginally tumorigenic human prostate epithelial cell line, when administered with either prostate or bone fibroblasts, were promoted to grow and disseminate in castrated hosts. Through this manipulation, we successfully developed a series of human prostate LNCaP sublines that share a cell lineage relationship with the parental LNCaP cells. The sublines C4-2 and C4-2B acquired both androgen independence and tumorigenicity as well as metastatic potential to the skeleton. We have observed that these cell lines, when administered either subcutaneously or orthotopically, can progress from their primary site of growth to the lymph node and eventually to the bone with a significant latent period (average 6.8 months) [8]. To understand whether this model may be used efficiently for assessing prostate cancer bone interaction, we have recently inoculated either parental LNCaP or its sublines directly to the skeleton and observed the growth of prostate cancer cells in the bone [10, Wu et al. Unpublished observation]. In a series of studies, we have demonstrated that although LNCaP failed to grow in the bone, C4-2 and C4-2B were observed to grow successfully in the skeleton. We have observed time-dependent increases of PSA in serum elaborated by prostate cancer cells grown in the bone. In concordance with the data obtained from patients, we have seen that despite elevation of serum PSA following prostate cancer growth in the bone, circulating tumor cells in serum had no correlation with serum PSA. This observation agrees with the clinical observation that serum PSA bears no relationship to circulating prostate cancer cells in the blood [11]. In a recent publication [4], we demonstrated that progression-associated prostate cancer genetic modifications can be observed upon prostate cancer progression. Stage- and progression-specific genetic alterations were induced in LNCaP and its lineage-derived cell lines, suggesting that genetic changes of prostate cancer cells may be subjected to epigenetic cues from the surrounding stromal cells.

ARCaP represents an androgen-repressed human prostate cancer cell line derived from the ascites fluid of a man with metastatic prostate cancer. The patient was considered to have rapid progressive and invasive disease. Results obtained from experimental studies of ARCaP cells in mice closely mimicked the clinical characteristics of the ARCaP cells in the patient [12]. Table 1 lists the characteristics of ARCaP cells in the patient and the behaviors of ARCaP cells in athymic mice.

Table 1 ARCaP cell phenotypes in mice closely mimicked behaviors in the patient

	in mice	In patient
Gene expression		
PSA	+	+
AR	+	+
Neuroendocrine factors	+	+
Metalloproteinases	+	ND
Metastasis		
Bone	+	+
Kidney	+	+
Lung	+	+
Pancreas	+	ND
Adrenal	+	ND
Ascites fluid	+	+

ND: Not determined.

ARCaP cells expressed low levels of androgen receptor and PSA, and appeared to be invasive and metastatic when inoculated orthotopically in the animal [12]. As in the original patient, ARCaP cells in nude mice metastasized to bone, kidney, lung, pancreas, and adrenal glands with high levels of metalloproteinase expression. This cell line may represent the further progression of prostate cancer from its androgen-refractory or -insensitive state toward androgen repression [2, 12]. Androgen-repressed human prostate cancer can be viewed as a positive responder to intermittent androgen therapy and potentially sensitive to flutamide withdrawal therapy. In both cases, we hypothesized that circulating androgen, or estrogen derived from aromatization of either endogenous or exogenous androgens, and growth factor-receptor pathways that may be activated upon flutamide withdrawal can significantly attenuate the growth of prostate cancer cells that exhibit androgen- and estrogen-repressed phenotypes. ARCaP cells failed to amplify their androgen receptors and appear to have no structural mutations at either DNA- or steroid-binding domains [Zhang, et al. Upublished oberservation]. This suggests the following three mechanisms of androgen suppression mediated by the androgen receptor: (1) androgen receptor mutation at exon 1, (2) interaction between androgen receptor and other interactive proteins, and (3) influencing the growth of prostate cancer cells by epigenetic factors, such as factors secreted by host microenvironment and stroma which affect tumor angiogenesis and attenuate host immune responses to the growing tumors. Presently, we do not know the mechanisms for androgen-induced repression of prostate cancer growth and metastasis.

Fig. 1 Models developed to study human prostate cancer progression

Cell line	LNCaP		C4	C4-2/C4-2 B		ARCaP
Phenotype	Androgen-dependent	⟶	Androgen-independent		----→	Androgen-repressed
Androgen stimulation	+++		+	−		−
Androgen repression	−		−	+		++
AR expression	++		++	+		+
PSA expression	+		++	+++		+
Invasion/metastasis	−		−	+		++

Fig. 1 Progression of human prostate cancer from an androgen-dependent to an androgen-repressed state.

The parental LNCaP and its lineage-derived cell lines C4, C4-2, and C4-2 B (and C4-2 B sublines) are a series of cell lines with varied androgen dependency and metastatic potential. Upon culturing LNCaP cells under androgen-deprived conditions for a long period, the repression of a LNCaP cell clone by androgen was observed [Umekita *et al.*, 1996]; the repressed LNCaP cell clone can grow in castrated males under the support of Matrigel, a reconstituted basement membrane. These results are consistent with the concept that androgen-responsive LNCaP cells can be "triggered" to exhibit androgen repression under the influence of altered cell culture conditions. ARCaP, a human prostate cancer cell line derived from an orchiectomized patient with highly invasive disease, represents the end stage of prostate cancer. Based upon these observations, we propose that prostate cancer progressed from the androgen-dependent to the androgen-independent state through cancer cell-stromal interaction. By manipulating the growth conditions of the cancer cells (e. g., androgen deprivation), prostate cancer may progress further to become androgen and estrogen repressed. The repressed clone(s) may acquire additional invasive and metastic potential, which could be lethal to the host.

Gene therapy

In view of the rapid progress in understanding the molecular and biological aspects of prostate cancer, much effort has recently been devoted to exploring the possible uses of gene therapy to control prostate cancer growth, differentiation, apoptosis, and senescence. Our laboratory was among the first to use adenoviruses to deliver the therapeutic toxic gene thymidine kinase and tumor suppressors (p 53, p 21, and p 16) to eradicate the growth of androgen-independent prostate cancer tumors in experimental models [6, 3]. We observed that tumor suppressors p 53, p 21, and p 16 and the toxic gene thymidine kinase were effective in inhibiting the growth of androgen-independent human prostate cancer cells both *in vitro* and *in vivo*. Significant survival and much reduced tumorigenicity was found by the use of tumor suppressors delivered by a universal promoter CMV. In these studies, we have observed that among the tumor suppressors, p 53 appears to be the most potent inhibitor of prostate tumor growth. p 16 in turn had a greater tumor suppressor effect than p 21. Tumor suppressors appear to be more potent when inhibiting the growth of smaller rather than larger tumors [3]. These results, taken together, suggest that tumor suppressors and toxic genes can block growth, induce apoptosis, and cause cell cycle arrest of androgen-independent growth of human prostate cancer cells in animal models.

To improve the delivery and expression of therapeutic tumor suppressors or toxic genes in human prostate cancer cells, we have recently developed tissue-specific promoter-driven therapeutic genes and expressed them in localized and metastatic prostate cancers. In this effort, we have explored the possibility of using the PSA promoter as a delivery system to express toxic genes in androgen-dependent as well as androgen-independent LNCaP lineage-derived cell lines C4-2 and C4-2 B. This model system is believed to be relevant to human prostate cancer progression for the following reasons: (1) LNCaP and lineage-derived cell lines expressed PSA and PSMA. The growth of LNCaP is stimulated by dihydrotestosterone both *in vivo* and *in vitro*; and (2) the LNCaP-derived cell line C4-2 expressed high levels of PSA in the absence of testicular androgen [8, 10]. This cell line appears to be androgen-independent and exhibits metastatic potential when inoculated orthotopically [8].C4-2 cells were observed to metastasize from the primary tumor to lymph nodes, and eventually to distant bone sites with marked osteoblastic reactions. This model was chosen for our studies in part because of the expression of a useful human prostate cancer marker PSA; this mimics the human prostate cancer condition where advanced androgen-independent prostate cancer expressed high levels of PSA. Using this experimental model, we have clearly demonstrated that an adenovirus, Ad-PSA-TK, which contains a long PSA promoter enhancer (p 61), expressed high levels of TK activity and inhibited the growth of both PSA-producing androgen-dependent LNCaP and androgen-independent C4-2 cell lines *in vitro* and tumor growth *in vivo*. Similar Ad-PSA-TK, however, was ineffective in inhibiting the growth of a human transitional cell carcinoma WH growth *in vitro*. These observations clearly demonstrated that the PSA promoter has the advantage of direct delivery of therapeutic genes to responsive cells or tissues. One advantage of using the p 61 PSA promoter to deliver therapeutic genes to prostate cancers is that this promoter can elicit high transactivating activity for therapeutic toxic genes in androgen-refractory human prostate cancer without resorting to the administration of exogenous androgen. This strategy can overcome the controversial limitations of androgen administration to men with metastatic prostate cancer.

Another approach we have developed is the effective use of the osteocalcin (OC) promoter to drive the expression of therapeutic toxic genes in human prostate cancer.

Ad-OC-TK was shown to eradicate the growth of both rat and human osteosarcomas *in vivo* and *in vitro* [1, 5]. Ad-OC-TK was found to be expressed in a number of human calcified malignant diseases, including both androgen-dependent and androgen-independent prostate cancer, brain, mammary, and lung cancers. Using the OC promoter-driven therapeutic toxic gene delivery, we have accomplished the dual effect of eradicating the growth of prostate cancer cells while in the bone as well as their supporting osteoblasts. In an experimental model of human prostate cancer growth in bone, we demonstrated that the delivery of Ad-OC-TK together with the pro-drug ACV significantly impaired the growth of androgen-independent human prostate cancer cell lines C4-2 and PC-3 both *in vitro* and *in vivo*. These findings were supported further by the histomorphologic examinations of the tumors in the skeleton and x-rays of the bone lesions before and after treatment. Fig. 2 lists the general strategies of gene therapy for the treatment of prostate cancer metastasis.

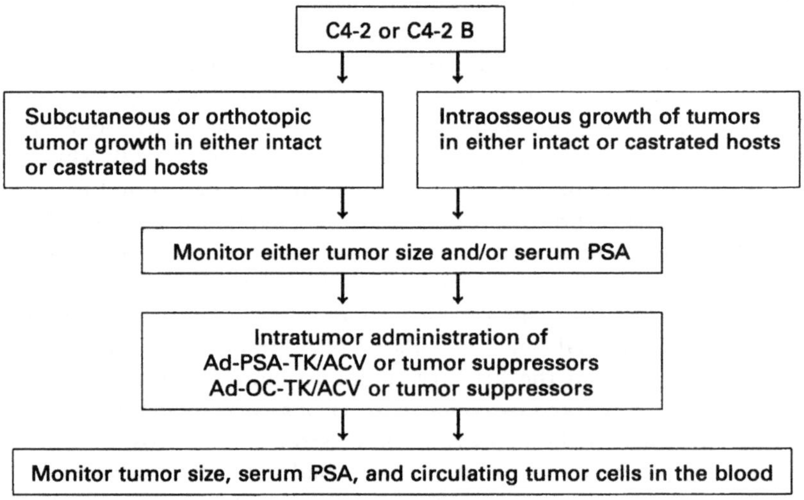

Fig. 2 Pre-clinical gene therapy protocol for the treatment of human prostate cancer metastasis.

Summary and future perspectives

We have developed a number of useful and unique animal models for the study of human prostate cancer growth and progression. These models are useful in analyzing markers that are indicative of prostate cancer progression and can be employed as a powerful system to study the potential genetic alterations associated with cancer progression and rational therapeutic approaches to block the growth of cancer cells *in vivo*. We have developed several classes of therapeutic adenoviral constructs, p 53, p 21, and p 16 tumor suppressors, and TK/ACV for bystander cell-kill, all of which have been observed to inhibit the growth of human prostate cancer experimental models. In addition, we have employed tissue- or tumor-specific promoters, including PSA and OC, to deliver and express high levels of therapeutic genes in tumor epithelium and/or its supporting stroma. This approach of targeting both the prostatic epithelium and the stroma will open new avenues for the treatment of prostate cancer bone metastasis.

Future directions in prostate cancer research rely upon the further investigation of the molecular mechanisms which signal prostate cancer growth differentiation apoptosis, and senescence. It is important, however, to focus on the clinically relevant targets, which unfortunately are poorly defined today. In view of the lack of knowledge of cancer biology and metastasis, an additional effort may lead to a more comprehensive understanding of the molecular pathways of cell signaling during prostate cancer progression. Close interaction between molecular/cell biologists and the surgeons of future generations ("molecular" surgeons) will lead to new strategies for the management of advanced prostate cancer.

Acknowledgments

The authors wish to acknowledge the contributions, both in concept and in experimental work, made by the following individuals: Song-Chu (Arthur) Ko, M. D.; Toshiro Shirakawa, M. D.; Tony T. Wu, M. D.; Thomas A. Gardner, M. D.; Hongquan Zhang, Ph. D.; and Akinobu Gotoh, M. D. The excellent secretarial and editorial support provided by Leilani Caven and Gary Mawyer for this manuscript is greatly appreciated. Financial contributions from RO1 CA64863, the CaP CURE Foundation, and the American Foundation of Urologic Diseases are gratefully appreciated.

References

[1] Cheon, J., S. C. Ko, T. A. Gardner, T. Shirakawa et al.: Osteosarcoma: Osteocalcin promoter based suicide gene therapy in combination with methotrexate in a murine osteosarcoma model. Cancer Gene Therapy *4* (1997) 359–365.

[2] Chung, L. W. K., H. Y. E. Zhau, T. T. Wu: Development of human prostate cancer models for chemoprevention and experimental therapeutics studies. J. Cell Biochem. (in press).

[3] Gotoh, C. Kao, S. C. Ko et al.: Cytotoxic effects of recombinant adenovirus p53 and cell-cycle regulator genes ($p21^{WAF17ciP1}$ and $p16^{INK4}$) in human prostate cancers. J. Urol. *158* (1997) 636–641.

[4] Hyytinen, E.-R., G. N. Thalmann, H. E. Zhau et al.: Genetic changes associated with the acquisition of androgen-independent growth, tumorigenicity and metastatic potential in a prostate cancer model. Brit. J. Cancer *75* (1997) 190–195.

[5] Ko, S. C., J. Cheon, C. Kao et al.: Osteocalcin promoter-based toxic gene therapy for the treatment of osteosarcoma in experimental models. Cancer Res. *56* (1996 a) 4614–1619.

[6] Ko, S. C., A. Gotoh, G. N. Thalmann, H. E. Zhau et al.: Molecular therapy with recombinant p53 adenovirus in an androgen-independent metastatic human prostate cancer model. Human Gene Therapy 7 (1996 b) 1683–1691.

[7] Parker, S. L., T. Tong, S. Bolden et al.: Cancer Statistics *47* (1997) 5–27.
[8] Thalmann, G. N., P. E. Anezinis, S.-M. Chang et al.: Androgen-independent cancer progression and bone metastasis in the LNCaP model of human prostate cancer. Cancer Res. *54* (1994) 2577–2581.
[9] Umekita, Y., R. A. Hiipakka, J. M. Kokontis et al.: Human prostate tumor growth in athymic mice: inhibition by androgens and stimulated finasteride. Proc. Nat'l. Acad. Sci. USA *93* (1996) 11802–11807.
[10] Wu, H. S., J. T. Hsieh, M. E. Gleave et al.: Derivation of androgen-independent human LNCaP prostatic cancer cell sublines: role of bone stromal cells. Int. J. Cancer *57* (1994) 406–412.
[11] Zhang, Y., C. D. Zippe, F. Van Lente et al.: Combined nested reverse transcription-PCR assay for prostate-specific antigen and prostate-specific membrane antigen in detecting circulating prostatic cells. Clin. Cancer Res. *3* (1997) 1215–1220.
[12] Zhau, H. Y. E., S.-M. Chang, B.-Q. Chen et al.: Androgen-repressed phenotype in human prostate cancer. Proc. Nat'l. Acad. Sci. USA *93* (1996) 15152–15157.

Tumor angiogenesis: the VEGF signalling system is a novel target for prostate cancer therapy

D. Marmé

Introduction

The growth of solid tumors and the formation of metastases are dependent on the formation of new blood vessels. Solid tumors have their origin in one genetically altered cell which, due to the accumulation of activated oncogenes and inactivated tumor suppressor genes, gains some growth advantage and expands to a small clone of a few millimeters in diameter. It is only after turning on the "angiogenic switch" that the tumor cell clone gains access to the vascular system. This hypothesis was proposed by Judah Folkman more than 25 years ago and has been substantially supported by recent discoveries which have unravelled the molecular and cellular processes involved in tumor angiogenesis [11]. Increasing knowledge about the various steps involved has led to the development of a number of anti-angiogenic strategies; some of these have even reached the clinic [7]. Recently, endogenous negative regulators of angiogenesis have been discovered and their involvement in the pathogenesis of metastasis has been suggested [4, 28]. Thus, it can be concluded that either the increased production of angiogenic factors or the decreased levels of negative regulators of angiogenesis, or both, are involved in the onset of tumor angiogenesis. This suggests that there are many opportunities for therapeutic intervention.

There is now quite convincing evidence indicating that the number of microvessels which can be detected in primary tumors is correlated with the subsequent appearance of metastases. Tumor angiogenesis has been reported to be a new prognostic tool in breast carcinoma, in oral cavity tumors and in invasive bladder carcinomas. However, it is not only the prognostic value of angiogenesis that makes it an interesting field for clinical-research: quantification of microvessels in the primary tumor might also be useful for selecting those patients who might benefit from an adjuvant therapy directed against the formation of new blood vessels or even leading to the destruction of newly formed blood vessels (for review, see [21]).

Among the various angiogenic factors, the vascular endothelial growth factor (VEGF) plays a pivotal role in tumor angiogenesis: It is expressed by most solid tumors; it is secreted; its receptors are preferentially expressed in vessels lining and penetrating the tumors; and the receptors are expressed almost exclusively on endothelial cells. This latter point is of particular interest because it suggests that there may be a sufficiently large therapeutic window for successful cancer therapy without the considerable side-effects of conventional cytostatic approaches. This brief review compiles evidence supporting the concept of VEGF-mediated tumor angiogenesis as a promising target for a novel tumor therapy.

According to our present understanding of cancerogenesis, each tumor originates from a single cell which has been transformed by one or more genetic events (Fig. 1). This includes the activation of oncogenes as well as the inactivation of tumor suppressor genes. These cells can grow out to form a cell clone of only a few millimeters in size before the supply of nutrients becomes limited. The tumor clones can remain at this stage for months or years, unless they are eliminated by the immune system or the so-called angiogenic switch is turned on. At the cellular level, the angiogenic switch is defined by the onset of expression and secretion of angiogenic factors. Assuming the presence of negative regulators of angiogenesis, the angiogenic factors have to override the delicate balance between positive and negative regulators.

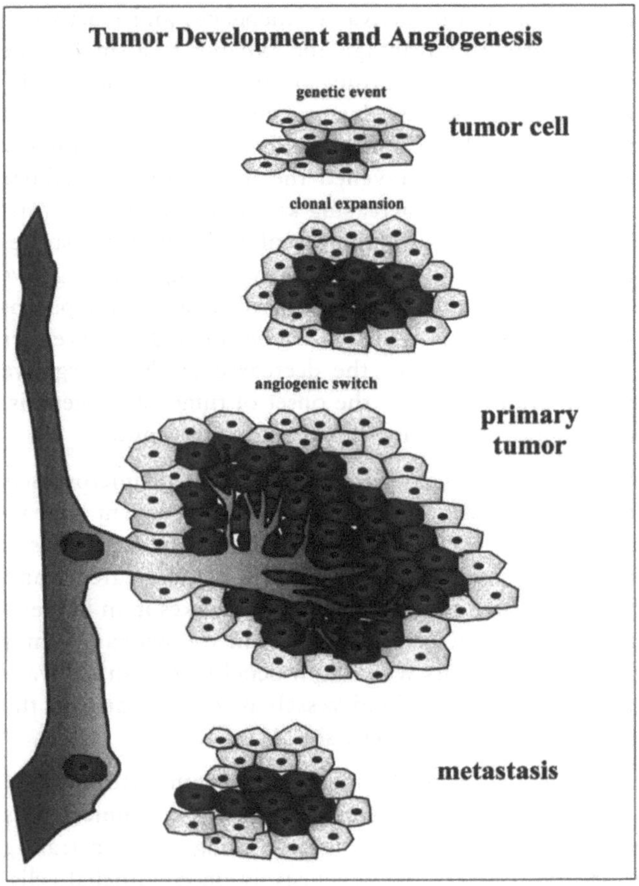

Fig. 1 Tumor development and angiogenesis.

If sufficient amounts of the secreted angiogenic factors reach the neighboring capillaries of the vasculature by diffusion they initiate the sprouting of new blood vessels from pre-existing ones.

Tumor development and vascularization

One of the first and most elegant investigations of the correlation between the onset of angiogenesis and tumor growth was carried out by Folkman and co-workers [12]. Using a transgenic mouse model, expressing an oncogene in the beta cells of pancreatic islets, they were able to follow the induction of angiogenesis during the transition from hypoplasia to neoplasia. It turned out that angiogenic activity first appears in a subset of hypoplastic islets before the onset of tumor formation. This study has stimulated a number of investigators to elucidate the relationship between angiogenesis and tumor growth and metastasis in human cancer. Using color doppler sonography to detect blood flow in pulpable and/or mammographically detected breast tumors Grischke et al. [15] were able to show that blood flow in the peritumoral region was significantly higher in malignant than in benign breast lesions. Quantification of microvessels revealed that the density of microvessels correlated with tumor size. Weidner and co-workers (for review see [38]) could show that a statistically significant correlation exists between the incidence of metastases and microvessel density of primary invasive breast carcinoma. Furthermore, there was a significant association of microvessel density with overall survival and relapse-free survival in all patients including node-negative and node-positive subsets [40].

Recently, it has been demonstrated for prostatic carcinoma that tumor angiogenesis is related to tumor growth and to metastatic potential [36]. The same group also presented evidence for a correlation of vascularization and tumor volume in latent prostatic carcinoma [13]. An additional study showed that the density of microvessels within prostate carcinomas was almost doubled in tumors from patients with metastases as compared to patients without metastases [39]. To investigate angiogenesis as a predictor of prognosis in invasive bladder carcinoma Dickinson et al. [5] could show that tumor vascularity is significantly correlated with patient survival.

From these studies it is evident that tumor growth as well as the formation of metastases is strongly related to the vascularization of the solid primary tumor. In addition, these findings suggest that microvessel density is a valuable prognostic factor.

The vascular endothelial growth factor VEGF

The vascular endothelial growth factor (VEGF) was recently identified as a secreted endothelial cell specific mitogen which is capable of stimulating angiogenesis *in vivo* (for review see [21]). Molecular cloning has revealed at least 4 different molecular isoforms having 121, 165, 189, and 206 amino acids, respectively. All 4 isoforms are encoded by a single gene and are generated by alternative splicing. Transcripts, encoding the 3 shorter forms, $VEGF_{121}$, $VEGF_{165}$, $VEGF_{189}$, are detected in the majority of tumor cells and tumor tissue expressing the VEGF gene. In contrast, $VEGF_{206}$ is a very rare form so far identified only in a human fetal liver cDNA library. All VEGF isoforms have a leader sequence and thus can readily leave the cells. The shortest form, $VEGF_{121}$, fails to bind heparin in contrast to $VEGF_{165}$ and $VEGF_{189}$. This reflects also their bioavailability with $VEGF_{121}$ being secreted as a freely soluble protein. The two other forms are significantly retained by the cell surface or the extracellular matrix. All 4 isoforms contain glycosylation sites which are essential for efficient secretion but not for biological activity. VEGF is active as a homodimeric molecule. $VEGF_{121}$, $VEGF_{165}$, and $VEGF_{189}$ have been produced recombinantly in insect cells [8]; the shortest form, $VEGF_{121}$, has been produced in Saccharomyces cervesiae [22]; and $VEGF_{165}$ was expressed successfully in E.coli [31]. All recombinant forms have been reported to be biologically active.

VEGF is a potent mitogen for vascular endothelial cells but it is apparently devoid of appreciable mitogenic activity for other cells. VEGF also promotes the migration of endothelial cells, their invasion into collagen gels, and the formation of tube-like structures. Recently it was demonstrated that VEGF binds to and induces migration of human monocytes [1]. It has also been shown that it specifically induces the formation and growth of new blood vessels in the chick chorioallantoic membrane [23]. Furthermore, it induces vascular permeability. This additional function of VEGF might be implicated in the mechanisms by which tumor cells enter and leave the blood stream and by which ascites is formed (Fig. 1).

Control of VEGF expression

The elucidation of the molecular mechanisms of the angiogenic switch by which tumor cells turn on the expression of angiogenic factors, such as VEGF, is crucial to the understanding of the whole process of tumor angiogenesis. According to the available experimental evidence, it is apparent that there is no single switch but rather a number of mechanisms which in combination cause the tumor cells to secrete sufficient amounts of VEGF to initiate and promote angiogenesis (Fig. 2).

As has been well established, tumor cells express various receptors for growth factors sometimes at very high levels. Therefore, it was interesting to see whether the activation of these growth factor receptors results in an enhanced expression of VEGF. It could be shown that platelet derived growth factor (PDGF) enhances the transcription of the VEGF gene considerably [9, 33]. Similar data have also been reported for EGF [14] and TGFβ [2, 6]. For PDGF it was also demonstrated that at least part of the effect was mediated by protein kinase C [9]. This is of particular interest because it has also been reported that tumor promoting phorbol esters which activate protein kinase C are able to induce angiogenesis in vitro [9] and that this induction could be inhibited by staurosporine, a potent protein kinase C inhibitor [27].

Enhanced VEGF gene expression was also observed upon stimulation of steroid hormone receptors. Estrogen as well as progesterone were able to stimulate VEGF mRNA to considerable levels [3]. Prostaglandins E_1 and E_2 which are potent stimulators of bone formation, a process strongly dependent on angiogenesis, were found to increase VEGF mRNA and protein levels in an osteoblastic cell line. This effect was suppressed by cAMP antagonists indicating the involvement of cyclic AMP in the

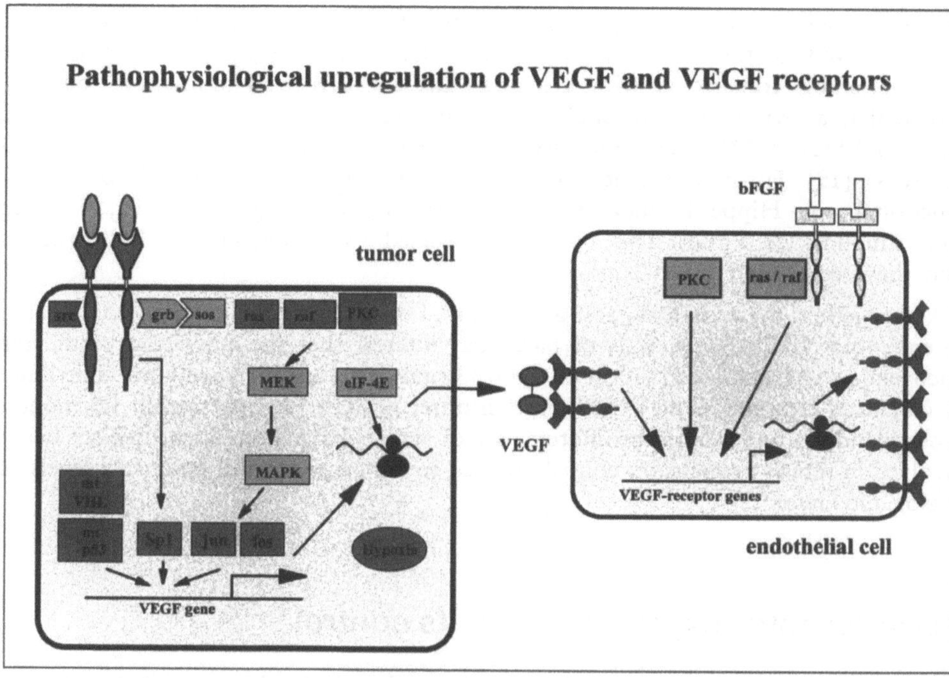

Fig. 2 Pathophysiological upregulation of VEGF and VEGF receptors.

process. Furthermore, the effect of prostaglandin E_2 was potentiated by retinoic acid and inhibited by dexamethasone [17].

Considerable VEGF expression was located close to necrotic areas of human glioblastomas [28 a]. This suggests that hypoxia has a promoting effect of VEGF expression. Using a glioblastoma cell line, Eli Keshet's group could demonstrate that exposure to hypoxia led to a dramatic increase of VEGF mRNA which returned to background when normal oxygen supply was resumed [30]. These findings have been confirmed in malignant as well as in normal cultured cells [10, 18, 25]. Concerning the underlying mechanism, hypoxia seems to increase VEGF mRNA stability rather than to substantially induce transcription. This is suggested by data obtained using a 2.068 base pair long VEGF promoter fragment fused to a luciferase reporter gene. This construct responded to PKC activating phorbol esters and PDGF but not to hypoxia. Furthermore, inhibition of transcription by actinomycin D did not influence the VEGF mRNA levels produced under hypoxia [10]. Interestingly, by using the same VEGF promoter/luciferase construct, it could be shown that the luciferase induction by TPA and PDGF was completely abolished in the presence of dexamethasone. This suggests the participation of functional AP-1 transcription factor [10].

The activated ras oncogene product and the inactivated p 53 tumor suppressor gene product are involved in the generation and progression of many human tumors. Therefore, it was of interest to investigate the potential involvement of the respective gene products in the regulation of VEGF gene expression. It turned out that $p\,21^{v\text{-}ras}$

causes a constitutive transcriptional activity of the VEGF gene. Similar effects were observed for the activated raf gene product [16]. Using a NIH 3T3 fibroblast cell line, stably transfected with a temperature sensitive p53 mutant, it could be shown that a switch from normal p53 to mutated p53 causes an increase in the levels of VEGF mRNA. This increase was potentiated by the activation of protein kinase C [19]. Furthermore, it was shown that renal carcinoma cells devoid of functional von Hippel-Lindau (VHL), tumor suppressor gene product, produce large amounts of VEGF. This could be reversed by reintroducing VHL wildtyp into these cells [32].

Taken together, these data suggest a substantial diversity of the mechanisms leading to enhanced VEGF levels. It is of particular interest that the activation of the ras oncogene as well the inactivation of the p53 tumor suppressor gene both contribute to enhanced expression of VEGF. Furthermore, growth factors, steroid hormones, prostaglandins, and tumor promoters, all of which have been identified as being involved in tumor promotion, contribute also to the generation of enhanced levels of angiogenic factor.

VEGF receptor expression and its control

Two high-affinity receptors for VEGF have been identified in humans: flt-1 and KDR (for review see [21]). The expression of both receptors *in vivo* appears to be restricted to endothelial cells and, most importantly, seems to be very low in healthy individuals. However, as it has been reported by many groups, overexpression of both receptors seems to be restricted to endothelial cells lining blood vessels adjacent to or growing into the tumors.

Thus far, evidence suggests that overexpression occurs always for both receptors. The mechanisms by which the VEGF receptors are up-regulated in tumor vessels is by far less understood as compared to the mechanisms controlling VEGF expression. Interestingly, tumor promoters and growth factors, such as bFGF and VEGF itself, are capable of enhancing VEGF receptor expression (B, Barleon and D. Marmé, unpublished data). However, it is reasonable to assume that signalling arising from tumor cells or stromal cells is involved by paracrine mechanisms in the up-regulation of VEGF-receptors in the vascular endothelium (Fig. 2). The identification of this factor or these factors will be crucial to the understanding of endothelial cell activation by angiogenic VEGF. Recently, it was shown that ex vivo perfusion of isolated lungs under hypoxic conditions caused a remarkable increase in mRNA for both human VEGF receptors flt-1 and KDR [34]. Whether this finding could be translated to the tumor situation is unclear.

Three recombinant isoforms, $VEGF_{121}$, $VEGF_{165}$, and $VEGF_{189}$ bind with similar affinities to both receptors and induce similar biological effects in human umbilical vein endothelial cells. Most of the observed biological responses are mediated by the VEGF receptor KDR. So far, it looks as if endothelial cell proliferation involves KDR but not flt-1 [36]. These data suggest that different biological responses might be generated by the two different VEGF receptors rather than by the different VEGF isoforms, however more work is needed to clearly attribute distinct VEGF receptors to distinct VEGF-mediated effects.

The VEGF/VEGF receptor system as a therapeutic target

The most intruguing question is, "could interference with the VEGF signalling system inhibit VEGF-mediated angiogenesis, tumor growth and metastasis"? Four lines of evidence support the view that interruption of the VEGF signalling pathway is indeed an effective strategy for inhibiting tumor progression and metastasis.

- First, three different VEGF-expressing human tumor cell lines were injected into nude mice. Animals received various doses of a monoclonal antibody against VEGF. Tumor progression from all three cell lines was inhibited by anti-VEGF antibodies. The antibodies had no effect on *in vitro* growth of the tumor cells, both on plastic and in soft agar. Furthermore, the density of vascular structures was decreased in anti-VEGF antibody-treated tumors when compared to controls [20].
- Second, functional inhibition of the mouse VEGF receptor, flk-1, the mouse homologue of human KDR, by the use of a dominant-negative mutant lacking protein tyrosine kinase, leads to significant inhibition of tumor growth and vascularisation. This effect was obtained either by co-injecting a cell line producing retroviruses encoding the dominant-negative mutant together with the glioblastoma cell line, C6, or by local injection of retroviral supernatants [24].
- Third, rat C6 glioma cells secreting VEGF were transfected with a eukaryotic vector expressing VEGF in antisense. Tumors arising after inoculating nude mice with these cells show reduced numbers of blood vessels, a higher degree of necrosis and inhibited growth rates [29].
- Fourth, using a VEGF-expressing human colon carcinoma cell line which has been shown to exhibit a high potential for liver metastasis, it was shown that treatment with monoclonal anti-VEGF antibody caused a dramatic reduction in the number and size of liver metastases [37].

From these data, it becomes apparent that an anti-VEGF therapeutic strategy could be successful in interfering, at least in part, with the vascularisation and growth of primary tumors as well as with the spread and outgrowth of metastases.

Conclusion

The implication of VEGF in tumor development and malignant progression has been evaluated to a point where it seems promising to develop anti-angiogenic strategy based on interference with the VEGF signalling system. A few targets in the chain of events seem to be especially promising:

- First, elimination of VEGF by antibodies or soluble, extracellular VEGF receptor domains.
- Second, interference with VEGF receptor-mediated signalling into endothelial cells, e. g., VEGF receptor antagonism, VEGF receptor dimerisation antagonism, VEGF receptor tyrosine kinase inhibitors.
- Third, since VEGF receptors occur primarily in endothelial cells and are overexpressed in the diseased area, this unique feature could afford specific targeting of VEGF-coupled inhibitors or toxins for therapeutic gene delivery.

As *in vitro* and *in vivo* evaluation studies of a therapeutic anti-VEGF strategy yield promising results, it is now important to develop clinical protocols for appropriate studies in humans.

References

[1] Barleon, B., A. Mantovani, D. Marmé et al.: Migration of human monocytes in response to vascular endothelial growth factor (VEGF) is mediated via the VEGF receptor flt-1. BLOOD *87* (1996) 3336–3343.

[2] Brogi, E., T. Wu, A. Namiki et al.: Indirect angiogenic cytokines upregulate VEGF and bFGF gene expression in vascular smooth muscle cells, whereas hypoxia upregulates VEGF expression only. Circulation *90* (1994) 649–652.

[3] Cullinan-Bove, K., R. D. Koos: Vascular endothelial growth factor/vascular permeability factor expression in the rat uterus: Rapid stimulation by estrogen correlates with estrogen-induced increases in uterine capillary permeability and growth. Endocrinology *133* (1993) 829–837.

[4] Dameron, K. M., O. V. Volpert, M. A. Tainsky et al.: Control of angiogenesis in fibroblasts by p53 regulation of thrombospondin-1. Science *265* (1994) 1582–1584.

[5] Dickinson, A. J., S. B. Fox, R. A. Persad et al.: Quantification of angiogenesis as an independent predictor of prognosis in invasive bladder carcinomas. Br. J. Urol. *74* (1994) 762–766.

[6] Dolecki, G. J., D. T. Connolly: Effects of a variety of cytokines and inducing agents on vascular permeability factor mRNA levels in U937 cells. Biochem. and Biophys. Res. Commun. *180* (1991) 572–578.

[7] Fan, T.-P. D., R. Jaggar, R. Bicknell: Controlling the vasculature: angiogenesis, anti-angiogenesis and vascular targeting of gene therapy. TiPS *16* (1995) 57–66.

[8] Fiebich, B. L., B. Jäger, C. Schöllmann et al.: Synthesis and assembly of functionally active human vascular endothelial growth factor homodimers in insect cells. Eur. J. Biochem. *211* (1993) 19–26.

[9] Finkenzeller, G., D. Marmé, H. A. Weich et al.: Platelet-derived growth factor-induced transcription of the vascular endothelial growth factor gene is mediated by protein kinase C. Cancer Res. *52* (1992) 4821–4823.

[10] Finkenzeller, G., A. Technau, D. Marmé: Hypoxia-induced transcription of the vascular endothelial growth factor gene is independent of functional AP-1 transcription factor. Biochem. and Biophys. Res. Commun. *208* (1995) 432–439.

[11] Folkman, J.: Angiogenesis in cancer, vascular, rheumatoid and other disease. Nature Medicine *1* (1995) 27–31.

[12] Folkman, J., Y. Shing: Angiogenesis. J. Biol. Chem. *267* (1992) 10931–10934.

[13] Furusato, M., S. Wakui, H. Sasaki et al.: Tumour angiogenesis in latent prostatic carcinoma. Br. J. Cancer *70* (1994) 1244–1246.

[14] Goldman, C. K., J. Kim, W.-L. Wong et al.: epidermal growth factor stimulates vascular endothelial growth factor production by human malignant glioma cells: A model of glioblastoma multiforme pathophysiology. Mol. Biol. Cell *4* (1993) 121–133.

[15] Grischke, E. M., M. Kaufmann, M. Eberlein-Gonska et al.: Angiogenesis as a diagnostic factor in primary breast cancer. Microvessel quantitation by stereological methods and correlation with color Doppler sonography. Onkologie *17* (1994) 35–42.

[16] Grugel, S., G. Finkenzeller, K. Weindel et al.: Both v-Ha-ras and v-raf stimulate expression of the vascular endothelial growth factor. J. Biol. Chem. *270* (1995) 25915–25919.

[17] Harada, S., J. A. Nagy, K. A. Sullivan et al.: Induction of vascular endothelial growth factor expression by prostaglandin E_2 and E_1 in osteoblasts. J. Clin. Invest. *93* (1994) 2490–2496.

[18] Iizuka, M., M. Yamauchi, K. Ando et al.: Quantitative RT-PCR assay detecting the transcriptional induction of vascular endothelial growth factor under hypoxia. Biochem. and Biophys. Res. Commun. *205* (1994) 1474–1480.

[19] Kieser, A., G. Brandner, H. A. Weich et al.: Mutant p53 potentiates protein kinase C induction of vascular endothelial growth factor expression. Oncogene *9* (1994) 963–969.

[20] Kim, K. J., B. Li, J. Winter et al.: Inhibition of vascular endothelial growth factor-induced angiogenesis suppresses tumour growth *in vivo*. Nature *362* (1993) 841–844.

[21] Kölch, W., G. Martiny-Baron, A. Kieser et al.: Regulation of the expression of VEGF/VPF and its receptors: role in tumorangiogenesis. Breast Cancer Research and Treatment *36* (1995) 139–155.

[22] Kondo, S., T. Matsumoto, Y. Yokoyama et al.: The shortest isoform of human vascular endothelial growth factor/vascular permeability factor ($VEGF/VPF_{121}$) produced by *Saccharomyces cerevisiae* promotes both angiogenesis and vascular permeability. Biochimia et Biophysica Acta *1243* (1995) 195–202.

[23] Kurz, H., S. Ambrosy, J. Wilting et al.: The proliferation pattern of capillary endothelial cells in chorioallantoic membrane development indicates local growth control, which is counteracted by vascular endothelial growth factor application. Developmental Dynamics *203* (1995) 174–186.

[24] Millauer, B., K. L. Shawver, K. H. Plate et al.: Glioblastoma growth inhibited *in vivo* by a dominant-negative Flk-1 mutant. Nature *367* (1994) 576–579.

[25] Minchenko, A., T. Bauer, S. Salceda et al.: Hypoxic stimulation of vascular endothelial growth factor expression *in vitro* and *in vivo*. Laboratory Investigation *71* (1994) 374–379.

[26] Morris, P. B., T. Hida, P. J. Blackshear et al.: Tumor-promoting phorbol esters induce angiogenesis in vivo. Cell. Physiol. *23* (1988) C318–C322.

[27] Oikawa, T., M. Shimamura, H. Ashino: Inhibition of angiogenesis by staurosporine, a potent protein kinase inhibitor. J. Antibiotics *45* (1992) 1155–1160.

[28] O'Reilly, M. S., L. Holmgren, Y. Shing et al.: Angiostatin: a novel angiogenesis inhibitor that mediates the suppression of metastases by a Lewis lung carcinoma. Cell *79* (1994) 315–328.

[28a] Plate, K. H., G. Breier, H. A. Weich et al.: Vascular endothelial growth is a potential tumour angiogenesis factor *in vivo*. Nature *359* (1992) 845–847.

[29] Saleh, M., A. F. Wilks: Presentation at the meeting on new cancer strategies: angiogenesis antagonists. Washington, D. C., April 1995, Abstract 49.

[30] Shweiki, D., A. Itin, D. Soffer et al.: Vascular endothelial growth factor induced by hypoxia may mediate hypoxia-initiated angiogenesis. Nature *359* (1992) 843–845.

[31] Siemeister, G., B. Schnurr, K. Mohrs et al.: Expression of biologically active isoforms of the tumor angiogenesis factor VEGF in E. coli. Biochem. Biophys. Res. Commun. 222 (1996) 249–255.

[32] Siemeister, G., K. Weindel, K. Mohrs et al.: Reversion of deregulated expression of vascular endothelial growth factor in human renal carcinoma cells by von Hippel-Lindau tumor suppressor protein. Cancer Res. *56* (1996) 2299–2301.

[33] Stauri, G. T., Hung, Y., Zachary, I. C. et al.: Hypoxia and platelet-derived growth factor-BB synergistically upregulate the expression of vascular endothelial growth factor in vascular smooth muscle cells. FEBS Letters *358* (1995) 311–315.

[34] Tuder, R. M., B. E. Flook, N. F. Voelkel: Increased gene expression for VEGF and the VEGF receptors *KDR/Flk* and *Flt* in lungs exposed to acute or to chronic hypoxia. J. Clin. Invest. *95* (1994) 1798–1807.

[35] Wakui, S., M. Furusato, H. Sasaki et al.: Tumor angiogenesis in prostatic carcinoma with and without bone metastases: a morphometric study. J. Pathol. *168* (1992) 257–262.

[36] Waltenberger, J., L. Claesson-Welsh, A. Siegbahn et al.: Different signal transduction properties of KDR and Flt 1, two receptors for vascular endothelial growth factor. J. Biol. Chem. *269* (1994) 26988–26995.

[37] Warren, R. A., H. Yuan, M. R. Matli et al.: Regulation by vascular endothelial growth factor of human colon cancer tumorigeneses in a mouse model of experimental liver metastasis. J. Clin. Invest. *95* (1995) 1789–1797.

[38] Weidner, N.: Tumor angiogenesis: Review of current applications in tumor prognostication. Seminars in Diagnostic Pathology *10* (1993) 302–313.

[39] Weidner, N., P. R. Carroll, J. Flax et al.: Tumor angiogenesis correlates with metastasis in invasive prostate carcinoma. Am. J. Pathol. *143* (1993) 401 –409.

[40] Weidner, N., J. Folkman, F. Pozza et al.: Tumor angiogenesis: A new significant and independent prognostic indicator in early-stage breast carcinoma. J. Nat. Cancer Inst. *84* (1992) 1875–1887.

Antimetastatic approaches to therapy of prostate cancer

R. Heicappell

Introduction

Metastasis to distant organs is the most devastating aspect of prostate cancer. Overall five years survival is more than 90% even if regional metastases are present [1]. In contrast, if distant metastases occur, 5-year survival is only 30%. Currently, the only generally accepted therapeutic approach to distant metastases is withdrawal of its major nutrient, testosterone. This approach is not curative, but releaves the symptoms of metastases and inhibits tumor growth for a variable period of time.

A rational approach to curative therapy of prostate cancer metastasis would consider the various aspects of prostate cancer progression such as detachment of cells from the parental tumor, invasion into lymphatics and blood vessels, and growth in target organs. New therapies based on interference with the various steps of prostate cancer metastasis are currently being developed using tumor cell lines in vitro and animal models of prostate cancer. Moreover, phase I and phase II clinical trials are currently being conducted with new antimetastatic drugs.

Crucial steps in prostate cancer metastasis

The process of cancer metastasis consists of a series of sequential steps each of which can be rate limiting because a failure at any of these steps eliminates the cells from the process.

After malignant transformation, tumor cells respond to growth signals to proliferate. If the tumor is less than 2 mm in diameter diffusion is sufficient to provide oxygen and nutrients. Any increase in size requires angionesis in response to angiogenic factors derived from the tumor or the host.

Detachment from the primary tumor and invasion into the blood and lymph circulation are the initial steps of tumor metastasis. Tumor cell invasion is facilitated by detachment from the primary tumor, increased cell mobility, and proteolytic enzymes such as type IV collagenase, cathepsins, and plasminogen activators.

The outcome of tumor metastasis is dependent on the interaction of the metastatic cell with the cells of the host in the circulation and different organ microenvironments.

In autopsy studies, prostate cancer metastases are found in the regional lymph nodes (87.3%), skeleton (78.6%), lungs (46.5%), liver (27.8%), and several other organs

[3]. The preferential sites of skeletal metastases at autopsy are the lumbar vetebrae (79%), the femur (78.5%), and the ilium (46.5%) [23].

It has been hypothesized that prostate cancer metastasis follows an anatomical route. Drainage of the prostate via Batson's plexus, a network of valveless venes that form extensive anastomoses with the venous systems of the vetebrae, pelvis, thorax, and brain. It can, however, not be excluded that metastatic prostate cancer cells reach their final destinations by entering the lymphatic circulation in the regional lymph nodes and the blood circulation via the thoracic duct. It is unlikeley that the mere presence of tumor cells in the circulation is sufficient to form distant metastasis.

Metastatic tumor cells undergo functional alterations, several of which have also been demonstrated in prostate cancer. Oncogenes such as c-erbB2 [2] are found overexpressed; metastasis suppressor genes such as KAI 1 [6] are downregulated during progression of prostate cancer. Metastatic tumor cells show increased motility and are capable of secreting serine proteases such as plasminogen activator and cystein proteases such as cathepsins. A detailed analysis of prostate cancer metastasis and the key factors involved in this process form the basis of new approaches to therapy.

Models for antimetastatic therapy in prostate cancer

There is no model for therapy of metastatic prostate cancer available that would simulate all features of metastatic human prostate cancer in experimental animals (Table 1).

Table 1 Models for human prostate cancer

Cell line	Source	Metastatic in nude mice	Androgen dependent	PSA production
PC-3	Bone metastasis	+	–	–
LnCaP	Lymphnode metastasis	–	+	+
DU-145	Brain metastasis	+	–	–

Some cell lines have low, if any, tumorigenicity in nude mice. Even if human prostate cancer cell lines grow and metastasize in nude mice, the major target of metastasis is not the skeleton as in human but the pelvic lymph nodes, visceral organs, or the lungs. From a clinical view, regional metastasis to pelvic lymphs nodes is less important than skeletal metastasis. Five year survival of the patients with lymph node metastasis is 93%. In contrast, patients with distant metastasis at time of initial diagnosis survival is only 30% [1]. Another important aspect, the transition of androgen-sensitive to androgen resistent prostate cancer and its implication for tumor progression have only recently been addressed in a newly developed human tumor model in SCID (severe combined immunodeficiency) mice [12].

It is therefore difficult to find appropriate animal models for development of antimetastatic therapy for prostate cancer.

Targets for antimetastatic therapy in prostate cancer

Potential targets for antimetastatic therapy are three major steps of tumor metastasis: Invasion into and extravasation from the circulation, circulating tumor cells, and growth of tumor cells in target organs.

For successful completion of each step of metastasis, factors inherent in the tumor cells as well as host factor have to be in a suitable combination in order to make metastasis an effective process. For each of the key steps of prostate cancer metastasis options for therapeutic interventions have been defined and introduced into in vitro or in vivo studies.

Inhibition of prostate cancer invasion and extravasation

Invasion into blood vessels and extravasation from the circulation are crucial steps in cancer metastasis. Tumor cells capable of invasion need to be equipped with adhesion molecules enabling them to adhere to the vascular endothelium or underlying structures and proteolytic enzymes for dissolution of the extracellular matrix.

Ideally, inhibition of tumor invasion would prevent the tumor cell from entering the blood stream. Thus, inhibition of molecules driving the tumor to invade vascular endothelium is a major rationale of antimetastatic therapy.

Anti-invasive activity of a given drug can be determined in vitro by the Boyden chamber assay. The Boyden chamber has two compartments that are separated by a polycarbonate filter with pores that allow cells to migrate from the upper to the lower compartment. The filter can be coated with proteins such as Matrigel^{TM}, a reconstituted basement membrane. Tumor cell invasion can then be quantitated by counting the cells that invade the filter in a given time.

In vivo, anti-invasive activity of a given drug can be determined by measuring the outcome of tumor progression, tumor metastasis.

Five principles have been applied to development of ant-invasive therapy of prostate cancer:

- Inhibition of proteolytic enzymes (e. g., metalloproteinases)
- Inhibition of neuroendocrine peptide hormones (e. g., bombesin)
- Chemotherapeutic drugs (e. g., taxol)
- Inhibition of signal transduction (e. g., carboxy-amido-triazol)
- Inhibition of regulatory proteins (e. g., uteroglobin)

Chemical inhibitors of matrix metalloproteinases (MMP) have been tested by Conway and coworkers [5] in vitro and in vivo in the rate prostate tumor model Mat-LyLu. After having demonstrated anti-invasive activity of these compounds in vitro, in vivo studies showed that the growth of primary tumors was inhibited at approximately 50% by MMP-antagonists. The number of lung metastases was, however, not lower in the treatment group as compared to untreated controls. Thus, at least in this model, antagonists to MMP are no effective antimetastatic drugs.

Urokinase type plasminogen activator (uPA) is a member of a family of serine proteases that converts plasminogen to plasmin and is thereby capable of clot lysis. Urinary plasminogen activator is believed to play a major role in tumor invasion and metastasis by breaking down structures of the extracellular matrix. In human prostate cancer uPA expression is inhibited by dexamethasone and stimulated by retinoic acid and epidermal growth factor [14].

Plasminogen activator inhibitor -1 (PAI-1) is an inhibitor of uPA expression. PAI-1 is often found in the microvasculature surrounding agressive tumors. Soff and coworkers [17] tested the hypothesis that expression of PAI-1 would inhibit prostate cancer invasion and metastasis. They examined the interaction between uPA and PAI-1 in a human prostate cancer cell line (PC-3 A) selected for high expression of uPA and low expression of PAI-1. This cell line was subsequently transfected with PAI-1 and examined in vitro and in vivo for its invasive and metastatic capability. In a limited number of experimental animals the authors showed that incidence of metastasis in lungs and liver was lower, if the tumor cells were transfected with PAI-1.

Uteroglobin is an immunomodulatory and inflammatory protein. It is a potent inhibitor of phospholipase A 2, a key enzyme in the arachidonic acid cascade. In prostate cancer, recombinant uteroglobin has been shown to inhibit invasion into reconstituted basement membrane towards chemotactical signals in several human tumor cell lines [13].

The neuroendocrine hormone bombesin reflects the presence of neuroendrocrine cells in prostate cancer. Hoosein and coworkers [11] showed that inhibition of bombesin by either antibody or bombesin antagonists inhibits the invasion of human prostate cancer cells into reconstituted extracellular matrix (Matrigel).

Matrigel invasion of human prostate cancer cell lines is inhibited by 60–70% after addition of carboxyamido-triazole (CAI). CAI is a calcium influx inhibitor that interferes with calcium sensitive signal transduction pathways. The effects of CAI are not anti-invasive only; the drug inhibits proliferation of several prostate cancer cell lines and suppresses the PSA secretion in LnCaP cells in a dose-dependent fashion [22]. Results from a phase I study have recently been reported [4]. A phase II study using CAI is currently being conducted by the National Cancer Institute of the USA.

Stearns and Wang [18] have examined the effects of taxol on invasion and metastasis of human prostate cancer in SCID mice using the PC-3 ML variant cell line that preferentially metastasizes to the lumbar vertebrae in SCID mice upon injection into the tail vein. In vitro, invasion into matrigel was inhibited in a dose- and time-dependent fashion. Inhibition of invasion was mainly due to inhibition of collagenase secretion. In vivo, after injection of taxol, no gross metastases could be detected in SCID mice whereas all controls had tumor in the lumbar vertebrae. Microscopically, however, tumor cells were detected in the bone marrow in all animals treated with taxol. Clearly, in this model, taxol is effective in reducing the number of gross metastases. Although initial clinical experience was disappointing in prostate cancer, new taxol based regimens are currently being evaluated in clinical trials in the US.

Inhibition of prostate tumor cell aggregation

It has been shown in experimental animals that the vast majority of tumor cells do not survive in the circulation [8]. Tumor cells surving in the circulation may form aggregates with other tumor cells and cell of the host such as leukocytes and/or platelets [8, 11]. Entrapment of tumor cell aggregates in the capillary bed is thought to be a major prerequisite for tumor metastasis. Consequently, dissolution of tumor cell aggregates would be one way to prevent metastasis. Platelet aggregating activity has been demonstrated in the human prostate cancer cell line PC-3. In vitro, purified components of snake venoms such as rhodostomin, trigramin, and the oligopeptide GRGDS have been shown to inhibit platelet activating activity of the prostate cancer cell line PC-3 in a dose-dependent fashion [19].

Inhibition of prostate cancer growth in target organs

The final outgrowth of metastasis depends in a suitable microenvironment in the target organ (for review, see [9]). In animal models, it has been shown that metastatic tumor cells do not grow in any given organ but exhibit preferences. Factors of the target organ supporting growth of tumor metastases include organ-specific fibroblasts and the supply with growth- and angiogenic factors. On the other hand, response to chemotherapy may be dependent on the expression of multidrug resistance genes which have been shown to be organ specific in some experimental systems [9].

In prostate cancer, interaction of metastatic tumor cells with bone has been evaluated in several experimental systems (for review, see [16]). Yet, new therapeutic approaches to bone metastases focus on new bone seeking radioisotopes [16].

Bisphoshonates are specific inhibitors of osteoclast-mediated bone resorption and have been shown to be effective in breast cancer and multiple myeloma. In osteoblastic metastases such as bone metastases from prostate cancer bisphosphonates provided a good palliation in some patients without affecting the growth of bone metastases [21]. So far, there are no convincing data supporting the hypothesis that bisphosphonates are drugs with specific activity on the growth of bone metastases from prostate cancer.

Perspectives of antimetastatic therapy in prostate cancer

Metastasis from prostate cancer is a major clinical problem and will remain so in the future. Some of the approaches to antimetastatic therapy will hopefully be translated into clinical studies. Currently, most of the evidence published is from experimental studies (Table 2). It will take much additional work before the approaches evaluated are feasible for the use in patients with metastatic prostate cancer.

Metastasis from prostate cancer is a multidimensional problem. It includes several aspects, such as genetic disposition, activation of oncogenes, down-regulation of tumor suppressor genes, dependence on hormones and growth factors, and interaction with host cells in the process of metastasis. In each of these fields much research is currently being done that will hopefully converge into new principles of therapy of metastatic prostate cancer in the future.

Table 2 Approaches to antimetastatic therapy of prostate cancer

Metastatic step involved	Principle	Drug/Therapy	Evidence	Remarks
Invasion	Inhibition of proteolytic enzymes – metalloproteinases	Antagonists to metallo-proteinases	Experimental (in vitro and in vivo)	Inactive in vivo [5]
	Inhibition of proteolytic enzymes – plasminogenactivator	Induction of plasminogen-activator inhibitor (PAI-1)	Experimental (in vitro and in vivo)	[17]
	Inhibition of neuro-endocrine peptide hormones	Bombesin antagonists	Experimental (in vitro)	[11]
	Chemotherapy	Taxol	Experimental (in vitro and in vivo)	[18]
	Inhibition of signal transduction	Carboxy-amido-triazol (CAI)	Clinical phase I	[4]
	Inhibition of regulatory proteins	Antibody to uteroglobin	Experimental (in vitro)	[13]
Circulation	Inhibition of tumor-derived platelet activating activity	Rhodostomin, trigramin, GRGDS	Experimental (in vitro)	[19]
Growth in target organ	Inhibition of osteoclast-mediated bone resorption	Clodronate	Clinical trial	No anti-metastatic activity in PCa [21]

References

[1] American Cancer Society, Cancer – Facts & Figures 1997, World Wide Web: http://www.cancer.org.

[2] Arai, Y., T. Yoshiki, O. Yoshida: C-erbB-2 oncoprotein: a potential biomarker of advanced prostate cancer. Prostate *30* (1997) 195–201.

[3] Berge, T., S. Lundberg: Cancer in Malmö 1958–1969; an autopsy study. Acta. Pathol. Microbiol. Scand. Suppl. *260* (1977) 1–235.

[4] Berlin, J., K. D. Tutsch, P. Hutson et al.: Phase I clinical and pharmacokinetic study of oral carboxyamidotriazole, a signal transduction inhibitior. J. Clin. Oncol. *15* (1997) 781–789.

[5] Conway, J. G., S. J. Trexler, J. A. Wakefield et al.: Effect of matrix metalloproteinase inhibitors on tumor growth and spontaneous metastasis. Clin. Exp. Metastasis *14* (1996) 115–124.

[6] Dong, J. T., H. Suzuki, S. S. Pin et al.: Down-regulation of the KAI 1 metastasis suppressor gene during the progression of human prostatic cancer infrequently involves gene mutation or allelic loss. Cancer Res. *56* (1996) 4387–4390.

[7] Fidler, I. J., C. Bucana: Mechanism of tumor cell resistance to lysis by syngenetic lymphocytes. Cancer Res. *37* (1977) 3945–3956.

[8] Fidler, I. J.: Modulation of the organ microenvironment for treatment of cancer metastasis. J. Nat.'l Cancer Inst. *87* (1995) 1588–1592.
[9] Fidler, I. J.: Metastasis: Quantitative analysis of distribution and fate of tumor emboli labeled with ^{125}I-5-iodo-2′-doexyuridine. J. Nat'l. Cancer Inst. *45* (1970) 773–782.
[10] Gasic, G. J.: Role of plasma, platelets and endothelial cells in tumor metastasis. Cancer Metastasis Reviews *3* (1984) 997–1004.
[11] Hoosein, N. M., C. J. Logothetis, L. W. K. Chung: Differential effects of peptide hormones bombesin, vasoactive intestinal polypeptide and somatostatin analog RC-160 on the invasive capacity of human prostatic carcinoma cells. J. Urol. *149* (1993) 1209–1213.
[12] Klein, K. A., R. E. Reiter, J. Redula et al.: Progression of metastatic human prostate cancer to androgen independence in immunodeficient SCID mice. Nature Medicine *3* (1997) 402–409.
[13] Leyton, J., M. J. Manyak, A. B. Mukherjee et al.: Recombinant human uteroglobin inhibits the in vitro invasiveness of human metastatic prostate tumor cells and the release of arachidonic acid stimulated by fibroblast-conditioned medium. Cancer Res. *54* (1994) 3969–3699.
[14] Liu, F. D., S. A. Rabbani: Induction of urinary plasminogen activator by retinoic acid results in increased invasiveness of human prostate cancer cells PC-3. Prostate 27 (1995) 269–276.
[15] Roth, B. J., B. Y. Yeap, G. Wilding et al.: Taxol in advanced, hormone-refractory carcinoma of the prostate. A phase II trial of the Eastern Cooperative Oncology Group. Cancer 72 (1993) 2457–2460.
[16] Scher, H. I., L. W. K. Chung: Bone metastases: improving the therapeutic index. Semin. Oncol. *21* (1994) 630–656.
[17] Soff, G. A., J. Sanderowitz, S. Gately et al.: Expression of plasminogen activator inhibitor type 1 by human prostate carcinoma cells inhibits primary tunor growth, tumor associated angiogenesis, and metastasis to lung and liver in an athymic mouse model. J. Clin. Invest. *96* (1995) 2563–2600.
[18] Stearns, M. E., M. Wang: Taxol blocks processes essential for prostate tumor cell (PC-3 ML) invasion and metastasis. Cancer Res. *52* (1992) 3376–3781.
[19] Swaim, M. W., H.-S. Chiang, T.-F. Huang: Characterisation of platelet aggregation induced by PC-3 human prostate adenocarcinoma cells and inhibited by venom peptides, trigramin, and rhodostomin. Eur. J. Cancer *32 A* (1996) 715–721.
[20] Uhr, J. W., R. H. Scheuermann, N. E. Street et al.: Cancer dormancy: Opportunities for new therapeutic approaches. Nature Medicine *3* (1997) 505–510.
[21] Vorreuther, R.: Bisphosphonates as an adjunct to palliative therapy of bone metastases from prostatic carcinoma. A pilot study on clodronate. Br. J. Urol. 72 (1993) 792–795.
[22] Wasilenko, W. J., A. J. Palad, K. D. Somers et al.: Effects of the calcium influx inhibitor carboxyamido-triazole on the proliferation and invasiveness of human prostate tumor cell lines. Int. J. Cancer *368* (1996) 259–264.
[23] Weiss, L., H. A. Gilbert (eds.): Bone metastasis, pp. 54–55. G. K. Hall Medical Publishers, Boston, MA.

Intermittent androgen suppression: rationale and clinical experience

M. Gleave, N. Bruchovsky, S. L. Goldenberg, P. Rennie

Abstract

Research on hormonal treatments of prostate cancer over the past years has focused on maximizing androgen ablation through combination therapy. Maximal androgen ablation prolongs time to androgen-independent (AI) progression by 3–6 months but increases treatment-related side-effects and expense. We have been studying the effects of continuous and intermittent androgen suppression (IAS) on tumour growth, apoptosis, and gene expression in tumour model systems and patients with prostate cancer. Observations from the LNCaP and Shionogi animal models suggest that progression to androgen-independence involves adaptive responses to androgen deprivation which may, in turn, be modulated by intermittent androgen replacement. Experiments in both model systems revealed that IAS almost tripled the time to AI progression.

Supported by these animal model observations, we have tested the efficacy of IAS therapy in a non-randomized group of 60 patients with a mean followup of 41 months using serum PSA as trigger points. Disease was clinically localized in 34 patients and metastatic in 26 patients. Mean initial serum PSA was 128 μg/L. Treatment was initiated with combined androgen blockade and continued for an average of 9 months. Only patients with PSA nadir levels below 4 μg/L are eligible for the IAS protocol. Medication was then withheld until the serum PSA increased to mean values between 10 and 20 μg/L. This cycle of treatment and no-treatment was repeated until the regulation of PSA became androgen independent. Twenty-nine men are currently in their 2nd cycle and 19 are in their 3rd or higher cycle. The mean time to PSA nadir during the first 3 cycles was 5 months. The first 2 cycles averaged 18 months in length with 45% of the time off therapy, while the 3rd cycle averaged 15.5 months. The off-treatment period in all cycles was associated with normalization of serum testosterone, an improvement in sense of well-being, and the recovery of libido and potency in the men who reported normal or near-normal sexual function before the start of therapy. Androgen-independent progression occurred in 9 of 26 stage D patients and 3 of 34 with localized disease after a mean followup of 43 months. Of 16 stage D2 patients, 8 remain on study with no stable disease. Eight patients have died, 2 from non-cancer related illness, with median time to progression and overall survival of 25 and 38.5 months, respectively.

Experimental and clinical data suggests that prostate cancer is amenable to control by IAS. This approach affords an improved quality of life when the patient is off-therapy and reduces toxicity and cost of treatment. Whether time to progression and survival is affected in a beneficial or adverse way is being studied in randomized, prospective protocols.

Background

No other treatment exists that equals or surpasses androgen ablation in controlling the growth of prostate cancer. Approximately 80% of prostate cancer patients achieve symptomatic and objective responses following androgen suppression and serum prostate specific antigen (PSA) levels decrease in almost all patients. Surgical or medical castration results in a median progression-free survival of 12 to 33 months and a median overall survival of 23 to 37 months in patients with stage D_2 disease [2, 6]. However, for reasons that remain unknown, the cell death process induced by androgen ablation by whatever means fails to eliminate the entire malignant cell population [4]. Another limitation of conventional androgen ablation is that it increases the rate of progression of prostate cancer to an androgen-independent state [4], and after a variable period of time averaging 24 months, the tumor inevitably recurs with increasing serum PSA levels and is characterized by androgen-independent growth. Over the past 20 years, most efforts have focused on maximizing the degree of androgen suppression therapy by combining agents that inhibit or block both testicular and adrenal androgens. However, maximal androgen ablation increases treatment-related side-effects and expense while prolonging time to androgen-independent progression by 3–6 months in most patients [6]. The paucity of information about cyclic hormonal therapy of breast and prostate malignancies can be attributed in part to the lack of suitable animal tumor models that regress predictably after castration and can be correctly termed hormone-dependent. Experimental animal and early clinical experience with intermittent androgen suppression (IAS) suggests that quality of life is improved and progression to androgen-independence may be delayed using reversible androgen suppression and PSA as the trigger point. IAS may offer a "way out" of the immediate vs. delayed treatment controversy, by balancing the benefits of immediate androgen ablation with reduced treatment-related side effects and expense.

Androgen dependence and independence

Androgen dependence is manifested by induction of apoptosis after androgen withdrawal in both normal and malignant tissues [3]. Androgen ablation has the double effect of triggering apoptosis and inhibiting DNA synthesis and cell proliferation. In the normal prostate, the cycle of androgen-induced cell growth and castration-induced apoptosis and regression can be continued through multiple cycles of androgen replacement and withdrawal. Data from animal studies and observations from the long term follow-up of Chinese eunuchs demonstrate that normal prostatic epithelial cells undergo apoptotic regression and do not develop the ability to regenerate and grow in an androgen-depleted environment [27]. In contrast, progression to androgen-independence nearly always occurs following androgen ablation therapy for patients with prostatic carcinoma. The inability of androgen withdrawal therapy to be curative may therefore result from pre-existing clones of androgen-independent cells in which the apoptotic process does not begin (clonal selection) [13, 14], or from the up-regulation of androgen-repressed adaptive mechanisms capable of aborting the apoptotic process in subpopulations of cells (adaptation) [4, 26].

Using the Dunning model, Isaacs et al. [13, 14] emphasizes the importance of genetic instability, which increases tumor heterogeneity by increasing the number of distinctly different clones of cells. Selective outgrowth of one or more androgen resistant clones may occur after androgen ablation, resulting in androgen-independent progression. However, Dunning tumors are androgen-sensitive and not androgen dependent, and apoptosis with tumor regression does not occur following castration. IAS may be harmful because it would result in stimulation of androgen-sensitive cells when testosterone levels increase. Indeed, in an earlier study, Russo et al. did not find IAS superior to castration alone in the Dunning model [20]. Maximal androgen ablation with the addition of cytotoxic agents that kill remaining androgen-idependent cells appears to be the most effective therapy in this model [13].

In contrast, observations from the Shionogi and LNCaP tumor models suggest that progression to androgen-independence is, in part, an adaptive response to androgen deprivation which involves induction of a new resistant phenotype that develops from scattered cells througout the tumor. The endpoint of androgen-independence in the Shionogi model is loss of castration-induced apoptosis and TRPM-2 gene expression [1, 4], while the endpoint for progression in the LNCaP model is androgen-independent regulation of PSA gene [7, 10]. Androgen withdrawal in mice bearing Shionogi tumors precipitates apoptosis and tumor regression in a highly reproducible manner; however, despite complete remissions, castration fails to result in cure and androgen-independent progression invariably occurs within 60 days following castration [4, 19]. Using an *in vivo* limiting dilution assay to determine the proportions of androgen-dependent and androgen-independent tumorigenic stem cells in parent and recurrent tumors, Bruchovsky et al. [4,] reported that the cells which survive androgen withdrawal may result from the ability of a small number of initially androgen-dependent stem cells to adapt to an androgen-depleted environment, possibly secondary to production of autocrine and paracrine growth factors which may replace androgens as primary growth stimulatory factors [22].

LNCaP is an androgen-sensitive, PSA-secreting, human prostate cancer cell line that is able to form tumors in male athymic mice under a variety of conditions [9], and provides accurate correlation between tumor volume and serum PSA levels *in vivo* [10]. Both androgens and tumor volume are important co-determinants of circulating PSA levels. Immediately after castration, serum PSA levels rapidly decrease by 80% and increase up to 20-fold following androgen supplementation; these changes in serum PSA occur without castration-induced tumor cell death or concomitant changes in tumor volume and reflect changes in androgen-regulated PSA gene expression. Despite decreasing by 80% after castration, LNCaP tumor cell mRNA PSA expression gradually returns to precastrate levels in the absence of testicular androgens beginning 4 weeks post-castration, heralding the onset of androgen-independent PSA gene regulation [28]. Further characterization of androgen-independent LNCaP cell lines produced a subline (C4-2) that metastasized to bone, and interestingly, the incidence of osseous metastases appeared higher in castrated than in intact male hosts, suggesting that continuous androgen suppression may facilitate development of AI osseous metastasis [12, 24].

The ability of benign or malignant prostate cells to undergo apoptosis and express PSA are acquired as a feature of differentiation of prostatic cells under the influence of androgens. In the absence of androgens, proliferating cells do not differentiate and cannot become preapoptotic again, which results in development of the androgen-independent phenotype and growth [3, 4]. Progression to androgen independence is likely genetically programmed (upregulation of androgen-repressed genes) but precipitated or directed by epigenetic factors (androgen ablation). In other words, androgen resistance may be a primary but quiescent property of some prostate cancer cells which is activated in response to androgen withdrawal. We hypothesize that if tumor cells which survive androgen withdrawal are forced into a normal pathway of differentiation by androgen replacement, then apoptotic potential might be restored and progression to androgen independence may be delayed. It follows that if androgens are replaced soon after regression of tumor, it should be possible to bring about repeated cycles of androgen-stimulated growth, differentiation and androgen-withdrawal regression of tumor.

Intermittent androgen suppression in animal models

Dunning R 3327 rat prostatic adenocarcinoma. Treatment of the Dunning R 3327 prostatic adenocarcinoma by cyclic therapy was studied by Trachtenberg [25] as a possible means of maintaining a population of hormonally dependent cells and preventing the emergence of an overwhelming number of hormonally independent cells. In his experiments, animals bearing the tumor were castrated and intermittently subjected to hormonal stimulation by means of indwelling silastic testosterone-filled implants. The growth of these tumors was compared to that of a castrate control group, a chronic implant group and an intact control group. No significant growth reduction with the intermittent stimulation group as compared to the implanted control or intact group was observed. It was concluded that intermittent hormonal therapy is inferior to early castration in preventing tumor growth.

The Dunning R 3327 prostatic adenocarcinoma was also used by Russo et al. [20] to study the effects of intermittent diethylstilbestrol diphosphate (DES) administration. This work was prompted by an exploratory clinical trial by Klotz et al. [5] which examined the use of intermittent DES therapy in advanced symptomatic prostate cancer to diminish the morbidity of standard endocrine therapy. Treatment of animals bearing the Dunning R 3327 tumor was started 60 days following implant and took the form of either intermittent high dose DES or intermittent low dose DES; in both cases the DES was administered in drinking water for one week, and off for three weeks. The growth of these tumors was compared to that of a castrate control group, and to two groups receiving either a low or high dose of DES administered continuously in drinking water. The continuous low or high doses of DES as well as the intermittent treatments were shown to reduce serum testosterone to castrate levels during the week of hormone administration; furthermore, serum testosterone recovered to control levels after withdrawal of DES. All rats exposed to DES had a tumor volume at death smaller than control or castrate rats. Despite this, a significant survival advantage from the time of randomization was achieved only in the castrate or high dose DES groups compared to control. Similarly, significant prolongation in tumor doubling time was achieved only in rats which had been castrated or had received high dose DES. Although the intermittent regimen was successful in checking tumor volume, it did not yield a survival advantage and in this respect was inferior to castration.

The foregoing experiments on the Dunning R 3327 prostatic adenocarcinoma draw attention to importance of using cyclic androgen deprivation only for the treatment of androgen-dependent cancer. If the tumor lacks the androgen-dependent phenotype and does not undergo apoptotic regression when androgens are withdrawn, intermittent therapy is unlikely to be beneficial. Because the Dunning R 3327 prostatic adenocarcinoma is androgen-sensitive but not androgen-dependent, the failure to demonstrate an improved outcome with intermittent hormone therapy is not an unexpected result.

Shionogi mouse carcinoma. The tumor best suited for studies on castration-induced cell death is the Shionogi carcinoma, a transplantable mouse mammary tumor which is androgen-dependent and closely mimics the clinical course of prostate cancer in response to treatment. The parent Shionogi tumor grows only in male mice, regresses almost completely after castration, and recurs with androgen-independent growth [4]. Because of its androgen-dependent behavior, the Shionogi tumor model is particularly useful in studying the mechanisms of androgen action, the phenotypic changes associated with apoptosis and androgen-independent progression, as well as therapeutic approaches to delay or avert tumor progression. Intermittent androgen suppression of the parent Shionogi carcinoma has been carried out experimentally by transplanting the tumor into a succession of male mice, each of which was castrated when the estimated tumor weight was approximately 3 g. After the tumor regressed to 30% of the original weight, it was transplanted into the next non-castrated male. This cycle of transplantation and castration-induced apoptosis was successfully repeated four times before growth became androgen-independent during the fifth cycle (Fig. 3) [1]. The mean time to androgen independence of 147 days compared to 51 days after one-time castration was in keeping with a retarding effect of cyclic therapy on tumor progression.

Some caution is required in the interpretation of these results since the procedure of transplanting a regressing tumor from a castrated to a non-castrated animal temporarily may reduce the rate of cell division and give rise to an apparent delay in progression. In fact, a transplantation-related effect on doubling time during the early phase of tumor regeneration would be technically difficult to rule out; however, from the time the tumor becomes palpable (0.5 g), no differences in tumour doubling times are evident during successive cycles [10]. It seems more likely that the lag period between tumor regression and regeneration is determined by the number of stem cells in the regressing tumor at the time when androgen replacements begins. Although castration reduces the number of viable tumor cells by about 1 logarithm, the effect on stem cells is disproportionately greater, i. e., 2–3 logarithms. In addition to tumor growth, other markers of androgen independent progression in this model include castration-induced apoptosis and TRPM-2 gene expression. Both these intermediate markers of androgen-dependence are maintained 3 times longer with IAS compared to continuous therapy, paralleling and supporting the changes in tumor volume.

LNCaP human prostate tumor model. As in human prostate cancer, serum PSA levels in this model are regulated by androgen and are directly proportional to tumor volume [10]. After castration, serum and tumor-cell PSA levels decrease up to 80% and remain suppressed for 3–4 weeks before increasing again (Fig. 1). With intermittent testosterone withdrawal and replacement in castrated mice bearing LNCaP tumors, a 3-fold delay in the development of androgen-independent regulation of the PSA gene was observed [21]. Serum PSA levels remained below precastrate levels after 60 days in 75% of the IAS group, but exceeded pre-castrate PSA after 28 days in 100% of the CAS group. By 15 weeks post castration, PSA levels in the continuous androgen suppression group had increased 7.0 fold above pre-castrate levels, compared to only

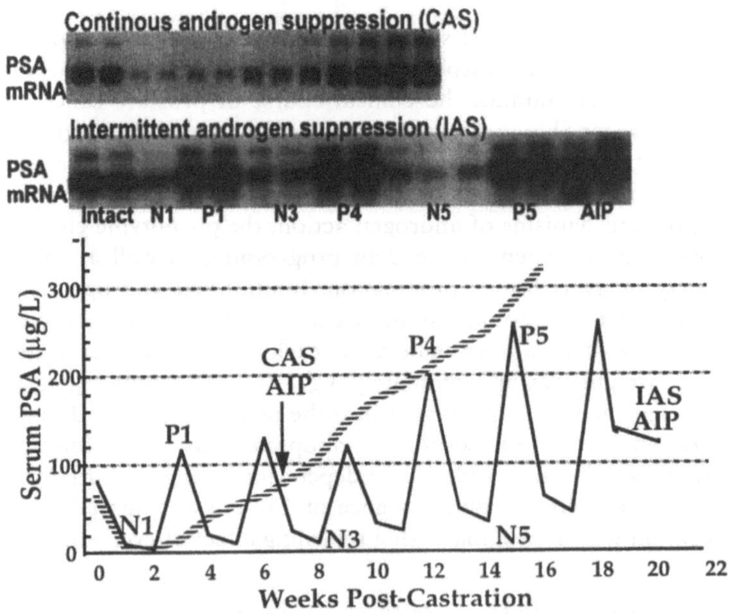

Fig. 1 The LNCaP tumor model uses androgen-regulated PSA gene expression as its endpoint of progression to androgen-independence (AIP). Serum PSA (graph) and PSA mRNA (Northern analysis) levels increase above precastrate levels by 4 weeks post-castration with continuous androgen suppression (CAS), but remained androgen-regulated and suppressed 15 weeks post-castration with IAS. By 15 weeks post castration, PSA levels with CAS had increased 7.0 fold above pre-castrate levels, compared to only 1.9 fold in the IAS therapy group.

1.9 fold in the IAS therapy group (Fig. 1). The success of IAS in these animal models supports the concept that progression to androgen-independence results from androgen-induced differentiation and/or down-regulation of androgen-repressed growth regulatory pathways.

Reversible androgen ablation therapy

With the introduction of anti-androgens such as cyproterone acetate and flutamide, and luteinizing hormone-releasing hormone agonists such as leuprolide acetate and goserelin acetate, new methods of androgen suppression have become available. These mimic the effects of orchiectomy by lowering the intranuclear concentration of dihydrotestosterone by 80% or more. Emphasis has been placed on the combined use of such agents [6] with little attention given to their reversibility of action, the significance of which is far-reaching. The potential for a full recovery from therapy makes it possible to alternate a patient between periods of treatment and no-treatment. Furthermore, serial serum PSA measurements permit accurate monitoring of disease activity and serves as trigger points for stopping and restarting therapy. When the patient is off-treatment, testicular function and serum testosterone levels return to normal slowly over a period of 8–14 weeks [1]. In response to this androgenic stimulus, atrophic cells are recruited into a normal pathway of differentiation where the risk of progression to androgen-independence is reduced. With the associated movement through the division cycle, the cells become pre-apoptotic again making it possible to repeat therapy.

Clinical observations

Reports of multiple regressions of hormone dependent malignancy are rare in the clinical and scientific literature. Stoll drew attention to the balance between inhibition and stimulation of the growth of breast cancer in an elderly woman who was treated intermittently with estrogen [23]. Regression of tumor occurred three times over a period of five years, after which an estrogen withdrawal response resulted in a forth remission. Bruchovsky described sequential responses of a pulmonary metastases of breast cancer in a female patient whose endocrine therapy consisted of radiation ovariectomy, administration of androgen, discontinuation of androgen and adrenalectomy respectively [5]. Regressions followed each procedure and the tumor size was successfully held in check. Experiments conduced by Noble focused on the effects of fractional hormone-replacement therapy on the progression of estrogen-dependent mammary tumors in male Nb rats [18]. In animals whose tumors achieved a stationary growth pattern with a reduced dose of estrogen, no autonomous change was observed. Noble also conducted similar studies on an adenocarcinoma of the dorsal prostate in Nb rats which grows only in estrogenized hosts [17]. Removal of estrogen from animals with growing tumors led to tumor regression; this was followed by progression and eventual regrowth of tumors that were autonomous. Replacement with lower doses of estrogen reduced the extent of regression and prevented autonomous change.

The intermittent regulation of serum testosterone levels for therapeutic purposes was first attempted with cyclic administration of estrogenic hormone [15]. Nineteen patients with advanced prostate cancer received diethylstilbestrol until a clinical response was clearly demonstrated and then it was withheld until symptoms recurred. One additional patient was treated with flutamide using a similar schedule. The mean duration of initial therapy was 30 months (range 2 to 70 months). Subjective improvement was noted in all patients during the first 3 months of treatment. When therapy was stopped, 12 of 20 patients relapsed after a mean interval of 8 months (range 1 to 24 months) and all subsequently responded to re-administration of drug. Therapy-induced impotency was reversed in 9 of 10 men within 3 months of the break in treatment. An improved quality of life was achieved owing to the reduced intake of diethylstilbestrol, and no adverse effects on survival were apparent.

In a more recent study, we reported preliminary results of IAS in a group of 47 patients with prostate cancer with a mean follow-up time of 125 weeks [11]. Sixty patients, 34 with clinically localized (17 untreated, 17 recurrent past local therapy) and 26 with metastatic (10 D1 and 16 D2) disease, have now been treated with mean follow-up of 41 months. Mean initial serum PSA was 128 µg/L. Treatment was initiated with combined androgen blockade and continued for an average of 9 months. Because prognosis is poor in patients who do not achieve normal PSA levels after androgen ablation, only patients with PSA nadir levels below 4 µg/L are eligible for the IAS protocol. Medication was then withheld until the serum PSA increased to mean values between 10 and 20 µg/L. This cycle of treatment and no-treatment was repeated until the regulation of PSA became androgen independent (Fig. 2).

Twenty-nine men are currently in their 2nd cycle and 19 are in their 3rd or higher cycle (Fig. 2). The mean time to PSA nadir during the first 3 cycles was 5 months. The first 2 cycles averaged 18 months in length with 45% of the time off therapy, while the 3rd cycle averaged 15.5 months. Serum testosterone returned to the normal range within 8 weeks (range: 1 to 26 weeks) of stopping treatment. The off-treatment period in all cycles was associated an improvement in sense of well-being, and the recovery of libido and potency in the men who reported normal or near-normal sexual function before the start of therapy.

Androgen-independent progression occurred in 9 of 26 stage D patients and 3 of 34 with localized disease after a mean followup of 43 months. Of 16 stage D 2 patients, 8 remain on study with stable disease. Eight patients have died, 2 from non cancer related illness, with median time to progression and overall survival of 25 and 38.5 months, respectively.

Observations from this preliminary study suggest that IAS does not have a negative impact on time to progression or survival, both of which are similar to continuous combined therapy. However, phase III randomized studies are required to accurately

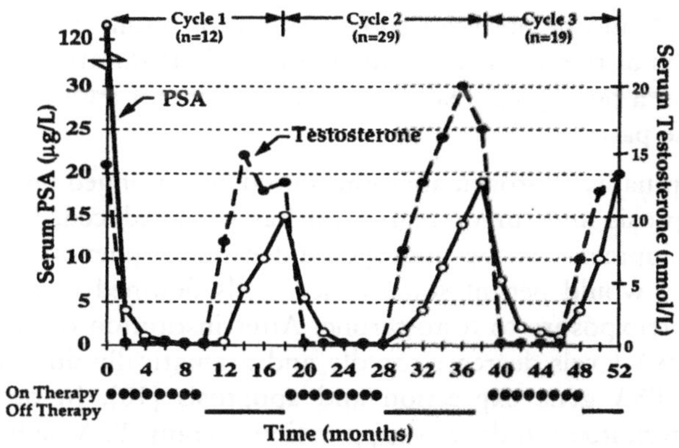

Fig. 2 Composite results on 60 patients with prostate cancer treated with intermittent androgen suppression, of whom 12 are in their 1st cycle of treatment, 29 in their 2nd cycle, and 19 in their third or higher cycle. Mean serum PSA at the start of the 1st, 2nd, and 3rd cycles was 128 µg/L, 14 µg/L, and 17.3 µg/L, respectively. The first 3 cycles averaged 18, 18, and 15.5 months in length, respectively.

assess the effects of intermittent treatment on these critical parameters. Intermittent androgen suppression does improve quality of life by permitting recovery of libido and potency, increasing energy levels and enhancing sense of well-being during off-treatment periods. The animal and preliminary clinical studies have helped identify groups of patients who are most likely to benefit from IAS, to determine optimal duration of treatment, and to suggest trigger points on when to re-start therapy again. The therapeutic strategy and trigger points in each situation is guided by serum PSA.

Candidates for intermittent androgen suppression. In theory, this approach should be suitable for the long-term management of inoperable, incompletely excised, or locally recurrent prostate cancer, especially after local recurrences following external beam ionizing radiation. The standard regimen developed for stage D 2 prostate cancer (Fig. 3) has been adapted for use in the treatment of men with initially localized prostate cancer who have recurred after either irradiation or radical prostatectomy

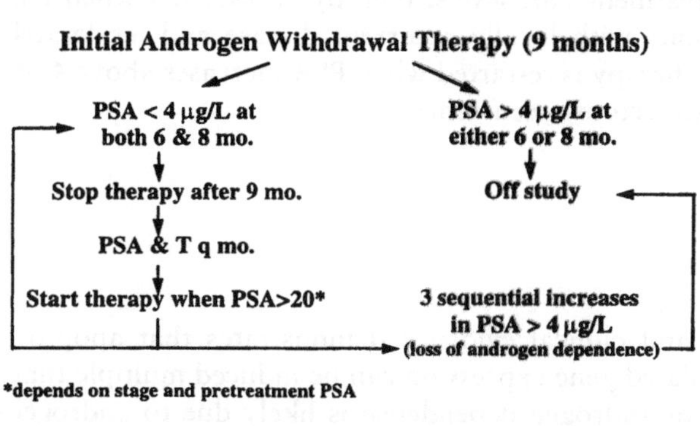

Fig. 3 Schema of intermittent androgen suppression theerapy for advanced prostate cancer.

as indicated by a rising serum PSA and/or positive biopsy. Failure of serum PSA to nadir below 4 µg/L after 6 months of therapy is associated with a short duration of response to androgen ablation and a poor prognosis [16], and intermittent androgen suppression is not offered to these patients.

Length of treatment cycle. Conceptually, androgen ablation should be continued until maximal castration-induced apoptosis and tumor regression has been induced, but stopped before constitutive development of the androgen-independent phenotype. Premature termination of therapy would permit some cancer cells destined to undergo apoptosis to survive upon reexposure to testosterone. After institution of androgen ablation therapy, serum PSA levels decrease rapidly and dramatically due to cessation of androgen-regulated PSA gene expression and apoptosis [10]. In our group of patients treated with intermittent androgen suppression therapy, PSA nadir and maximal soft tissue regression is not reached until eight months in many patients with advanced Stage C and D_2 disease [21]. Similarly, the duration of time necessary for PSA to reach its nadir in a group of patients with clinically confined disease after institution of neoadjuvant androgen withdrawal therapy is often longer than 6 months [8]. Using a lower limit of sensitivity of 0.2 µg/L, PSA nadir was reached in only 34% after 3 months, 60% after 5 months, 70% after 6 months, and 84% at 8 months. Serum PSA was still decreasing at 8 months in 14%. The percentage of patients with a nadir serum PSA level of $\leq$ 0.2 µg/L increased from 14% at 2 months, to 34% at 3 months, 56% at 5 months and 66% at 8 months. The initial rapid decrease in PSA results from cessation of androgen-regulated PSA synthesis and apoptosis, while the ongoing slower decline reflects decreasing tumor volume. Based on these observations, we now treat patients for 9 months prior to stopping therapy.

When to restart. The optimal time to reinstitute androgen ablation therapy remains undefined and empirical. Ideally, time off therapy should be long enough to permit normalization of libido and energy levels, and testosterone-induced tumor cell differentiation. Until more information is obtained about PSA milestones from clinical trials focusing on the subsets of patients indicated, the PSA thresholds and ceilings are regarded as tentative settings only. Trigger points are individualized and factors that are considered include pretreatment PSA levels, stage PSA velocity, presence of symptoms, and tolerance of androgen ablation therapy. In general, in patients with metastatic disease and high pretreatment PSA levels, therapy is restarted when PSA increases above 20 µg/L; in patients with locally recurrent disease and moderately elevated pretreatment PSA levels, therapy is restarted when PSA increases above 4 µg/L, or earlier for post-radical prostatectomy recurrences.

Conclusions

In summary, both experimental and clinical evidence demonstrates that apoptosis and modulation of androgen-regulated gene expression can be induced multiple times with IAS, and that maintenance of androgen-dependence is likely due to androgen-induced differentiation and/or suppression of alternative growth stimulatory pathways. The first 2 treatment cycles in our ongoing Phase II study averaged 18 months with 45% of the time off therapy, while the 3rd cycle averaged 15 months with 33% of the time off therapy. Time to progression and survival of 16 stage D 2 patients was similar to expected results with continuous therapy.

Available information about intermittent androgen blockade is still very limited and leaves several questions unanswered. The issue of survival is the most important and needs to be assessed in a randomized study of intermittent versus continuous androgen deprivation. The possibility of expanding the use of intermittent therapy to earlier stage disease should be examined in greater detail. Conceivably, the intermittent therapy option might become an alternative to radical prostatectomy or irradiation for the primary treatment of localized prostate cancer in men with life expectancies of less than 15 years. The prospect of improving the results of IAS by increasing the length and number of cycles is attractive. Prolongation of the length of each "off" treatment cycle may be possible with 5 α-reductase inhibitors like finasteride. Augmentation of intermittent therapy might be accomplished by the administration of cytotoxic drugs, differentiating agents, or gene therapies (e. g., antisense BCL2) at specific times during a cycle of treatment when the modality-of-choice would have its maximum effect. Finally, it is clear that the concept of IAS lends itself to other innovations as in the example of long-term neoadjuvant therapy prior to radical prostatectomy or irradiation [8].

References

[1] Akakura, K., N. Bruchovsky, S. L. Goldenberg et al.: Effects of intermittent androgen suppression on androgen-dependent tumours: apoptosis and serum prostate specific antigen. Cancer *71* (1993) 2782–2790.

[2] Bruchovsky, N.: Androgens and antiandrogens. In: Holland, E., Frei III, R. C. Bast et al.: Cancer Medicine. Third edition, pp. 884–896. Lea & Febiger, Philadelphia 1993.

[3] Bruchovsky, N., E. M. Brown, C. M. Coppin et al.: The endocrinology and treatment of prostate tumour progression. In: Current Concepts and Approaches to the Study of Prostate Cancer. D. S. Coffey, W. A. Gardner Jr., M. I. Resnick et al. (eds.): Progress in Clinical and Biological Research, pp. 348–387. Alan R. Liss. Inc., New York 1987.

[4] Bruchovsky, N., P. S. Rennie, A. J. Coldman et al.: Effects of androgen withdrawal on the stem cell composition of the Shionogi carcinoma. Cancer Res. *50* (1990) 2275–2282.

[5] Bruchovsky, N., P. S. Rennie, E. Van Doorn: Pathological growth of androgen-sensitive tissues resulting from latent actions of steroid hormones. J. Toxicol. Environ. Health *4* (1978) 391–408.

[6] Denis, L., G. P. Murphy: Overview of phase III trials on combined androgen treatment in patients with metastatic prostate cancer. Cancer 72 (Suppl.) (1993) 3888–3895.

[7] Gleave, M. E., M. Bowden, N. Bruchovsky, et al.: Predictors of time to androgen-independent progression in the LNCaP prostate tumor model. J. Urol. *151* (1994) 241.

[8] Gleave, M. E., S. L. Goldenberg, E. C. Jones et al.: Maximal biochemical and pathological downstaging requires 8 months of neoadjuvant hormonal therapy prior to radical prostatectomy. J. Urol. *155* (1996) 213–219.

[9] Gleave, M. E., J. T. Hsieh, C. A. Gao et al.: Acceleration of human prostate cancer growth in vivo by factors produced by prostate and bone fibroblasts. Cancer Res. *51* (1991) 3753.

[10] Gleave, M. E., J. T. Hsieh, H.-C. Wu et al.: Serum PSA levels in mice bearing human prostate LNCaP tumors are determined by tumor volume and endocrine and growth factors. Cancer Res. *52* (1992) 1598.

[11] Goldenberg, S. L., N. Bruchovsky, M. E. Gleave et al.: Intermittent androgen suppression in the treatment of prostate cancer: a preliminary report. Urology *45* (1995) 839–845.

[12] Hsieh, J. T., H.-C. Wu, M. E. Gleave et al.: Autocrine regulation of PSA gene expression in a human prostatic cancer (LNCaP) subline. Cancer Res. *53* (1993) 2852–2857.

[13] Isaacs, J. T.: The timing of androgen ablation therapy and/or chemotherapy in the treatment of prostatic cancer. Prostate *51* (1984) 1–17.

[14] Isaacs, J. T., N. Wake, D. S. Coffey et al.: Genetic instability coupled to clonal selection as a mechanism for progression in prostatic cancer. Cancer Res. *42* (1982) 2353–2361.

[15] Klotz, L. H., H. W. Herr, M. T. Morse et al.: Intermittent endocrine therapy for advanced prostate cancer. Cancer *58* (1986) 2546–2250.

[16] Miller, J. I., F. R. Ahmann, G. W. Drach et al.: The clinical usefulness of serum prostate specific antigen after hormonal therapy of metastatic prostate cancer. J. Urol. *147* (1992) 956–961.

[17] Noble, R. L.: Sex steroids as a cause of adenocarcinoma of the dorsal prostate in Nb rats, and their influence on the growth of transplants. Oncology *34* (1977) 138–141.

[18] Noble, R. L.: Hormonal control of growth and progression in tumors of Nb rats and a theory of action. Cancer Res. *37* (1977) 82–94.

[19] Rennie, P. S., N. Bruchovsky, R. Buttyan et al.: Gene expression during the early phases of regression of the androgen-dependent Shionogi mouse mammary carcinoma. Cancer Res. *48* (1988) 6309–6312.

[20] Russo, P., G. Liguori, W. D. W. Heston et al.: Effects of intermittent diethylstilbestrol diphosphate administration on the R 3327 rat prostatic carcinoma. Cancer Res. *47* (1987) 5967–5970.

[21] Sato, N., M. E. Gleave, S. L. Goldenberg et al.: Intermittent androgen suppression delays time to androgen-independent progression in the LNCaP prostate tumour model. J. Urol. *153* (1995) 282 A.

[22] Sato, N., Y. Watabe, H. Suzuki et al.: Progression of androgen-sensitive mouse tumour (Shionogi carcinoma 115) to androgen-insensitive tumor after long-term removal of testosterone. Jpn. J. Cancer Res. *84* (1993) 1300.

[23] Stoll, B. A.: Palliation by castration or by hormone administration. In: B. A. Stoll (ed.): Breast Cancer Management Early and Late, pp. 133–146. Year Book Medical Publishers, Inc., Chicago 1977.

[24] Thalmann, G. N., P. E. Anezinis, S. H. Chang et al.: Androgen-independent cancer progression and bone metastasis in the LNCaP Model of human prostate cancer. Cancer Res. *54* (1994) 2577.

[25] Trachtenberg, J.: Experimental treatment of prostatic cancer by intermittent hormonal therapy. J. Urol. *137* (1987) 785–788.

[26] Van Weerden, W. M.: Animal models in the study of progression of prostate and breast cancers to endocrine independency. In: P. M. J. J. Berns, J. C. Romijin, F. H. Schröder (eds.): Mechanisms of Progression to Hormone-Independent Growth of Breast and Prostate Cancer, pp. 55–70. Parthenon Publishers, New Jersey 1991.

[27] Wu, C. P., F. L. Gu: The prostate in eunochs. In: EORTC Genitourinary Group Monograph 10-Urologic Oncology: Reconstructive Surgery, Organ Preservation, and Restoration of Function, pp. 249–255. Wiley-Liss Inc., New York 1992.

[28] Wu, H.-C., J. T. Hsieh, M. E. Gleave et al.: Derivation of androgen-independent human LNCaP prostate cancer sublines: role of bone stromal cells. Int. J. Cancer *57* (1994) 406.

Resistance to drug therapy in cancer: gene therapy approaches to overcome anticancer drug resistance

Th. Licht

Abstract

Resistance of cancer cells to cytotoxic drug treatment is multicausative. Besides slow proliferation rates, a major reason for poor response to drug treatment of prostate cancer, specific mechanisms may contribute to chemoresistance. For instance, overexpression of P-glycoprotein, the product of the multidrug resistance (*MDR*1) gene confers resistance in cancers of various histological origins. Similar genes have been identified whose products function as drug efflux pumps or detoxifying enzymes. Molecular approaches to circumvent chemoresistance are aimed at inhibiting the function of proteins conferring chemoresistance, or interfering with transcription or translation of nucleic acids encoding these proteins.

Recent research has focused on a different approach to overcome drug resistance, employing chemoresistance genes such as *MDR*1 as useful tools for gene therapy of cancer. Transfer of this gene to hematopoietic cells was found to result in efficient expression, protecting them from the toxicity of various anticancer drugs. The latter approach may alleviate life-threatening side effects of chemotherapy such as leukocytopenia and thrombocytopenia, and increase the efficacy of chemotherapy by allowing safe dose-intensification.

Introduction

Prostate cancer displays resistance to a broad variety of cytotoxic drugs. Thus, the prognosis of patients suffering from hormone-refractory disease remains poor. Currently, no chemotherapeutic treatment can cure patients who fail to respond to hormone treatment. No single agent or combination chemotherapy have improved survival in randomized trials [36]. Several reasons may contribute to treatment failure in prostate cancer. Non-specific causes for chemoresistance include low proliferation rates as only actively dividing cells are vulnerable to the action of antimetabolites and drugs targeting the cytosceleton.

In addition, specific mechanisms for chemoresistance have been identified which may also contribute to a complex multidrug-resistant phenotype. For instance, P-glycoprotein, the product of the *MDR*1 (multidrug resistance) gene, is detectable at low levels in prostate cancer cell lines insensitive to hormone and cytotoxic treatment [51]. However, exposure to doxorubicin may induce expression of P-glycoprotein and chemoresistance in prostate cancer cells [4]. Induction of P-glycoprotein expression by chemotherapy is a well-known mechanism for the acquisition of drug resistance that has previously been reported in cancers of other histological origins [6, 7, 22, 29].

Other mechanisms of drug resistance include overexpression of the multidrug resistance-associated protein (*MRP*) gene [8, 9, 55] or the lung resistance-related protein (*LRP*) gene [42, 43] which confer resistance to multiple anticancer drugs. Different from the previously discussed mechanisms of drug resistance, mutations of the dihydrofolate reductase (*DHFR*) gene affect only antifolate drugs, e. g., methotrexate [13, 23]. An increasing number of molecular mechanisms is being discovered which render cancer cells insensitive to anticancer drugs. Moreover, several mechanism of drug resistance may simultaneously be expressed in cancers [53].

Drug resistance due to overexpression of P-glycoprotein has been extensively investigated for more than a decade (for reviews, see [15, 21]. P-glycoprotein-related multidrug resistance is well-characterized. Hence, this article will focus on strategies to circumvent P-glycoprotein related multidrug resistance which may serve as an example for other molecular mechanisms. Strategies to inhibit synthesis or function of P-glycoprotein, as well as more recent approaches to employ the *MDR*1 gene as a tool for gene therapy of cancer are being discussed. In spite of the widely held view that chemotherapy plays no role in prostate cancer, recent clinical studies have suggested that dose-intensification with administration of growth factors may change the natural course of prostate cancer [36, 46]. New strategies for circumvention of drug resistance might help to further improve the efficacy, or ameliorate the toxicity of chemotherapy.

Reversal of P-glycoprotein-mediated multidrug resistance by interference with synthesis or function of P-glycoprotein

Molecular strategies have been developed to interfere with synthesis of proteins confering resistance to anticancer drugs. This has been attempted with antisense oligonucleotides and ribozymes. Transcription of the human multidrug resistance gene has been blocked with antisense oligonucleotides [10, 14]. Similarly, reduction of *MRP* expression can be accomplished with the use of antisense oligonucleotides [48]. Like antisense oligonucleotides, ribozymes designed to cleave the *MDR*1 mRNA have the ability to restore chemosensitivity in multidrug resistant cells [27, 41]. Ribozymes are small RNA molecules displaying enzymatic activity to cut other mRNA molecules at specific sites. While efficient chemosensitizing has been demonstrated *in vitro*, however, the application of such strategies to clinical treatment of cancer has still to be awaited. Directing such molecules to cancer cells *in vivo* is elusive. Thus, new pharmacological vehicles have to be developed before these intriguing approaches can be transferred to the clinic.

Interest in circumvention of drug resistance has mainly been focused on inhibition of the function of P-glycoprotein. Verapamil, a calcium channel blocker, was one of the first compounds that was identified to reverse resistance to vincristine [52]. Soon thereafter, cyclosporine was found to be a modulator of resistance [45]. Both drugs are inhibitors of P-glycoprotein. Early clinical trials with verapamil revealed the feasibility of chemosensitization in certain patients. However, cardiac toxicity was unacceptable for routine anticancer treatment [40]. Cyclosporine caused side effects such as hypertension, nephrotoxicity and hyperbilirubinemia [54].

Due to its broad substrate specificity, P-glycoprotein interacts with several non-toxic compounds. Inhibitors of P-glycoprotein are structurally unrelated molecules, among them being steroid hormones and their antagonists [16, 39], neuroleptic agents such as thioxanthenes and phenothiazines [17, 18], antibiotics, and antifungal drugs [20, 28, 44]. Some inhibitors, e. g., the cyclosporine derivative PSC 833, are currently being investigated in clinical phase III trials.

The *MDR*1 gene in gene therapy of cancer

Although drug resistance genes are most frequently studied because they may cause (or are at least associated with) failure of anticancer treatment, such genes may conversely serve as useful tools for gene therapy of cancer. Transfer of chemoresistance genes to chemosensitive cells may render them resistant and protect them from the adverse effects of chemotherapy. This may be of particular interest for the hematopoietic system because most hematopoietic cells are highly susceptible to chemotherapy.

The effects of overexpression of P-glycoprotein in bone marrow were first studied in transgenic mice which expressed a human full-length *MDR*1 cDNA under the control of a chicken β-actin promoter [19, 33]. P-glycoprotein expression was detectable in bone marrow cells of all hematopoietic lineages. The normal function of bone marrow was not impaired, but several-fold higher doses of taxol and daunomycin could be administered safely as compared with normal animals of the respective background strains. Simultaneous administration of chemosensitizing agents like verapamil reversed chemoprotection in these mice, demonstrating the biological specificity of the effects due to *MDR*1 expression [34]. In subsequent experiments, bone marrow cells of *MDR*1-transgenic mice were transplanted into lethally irradiated recipient mice. It was demonstrated that the recipients were also protected from myelosuppression following chemotherapy [35].

Based on the investigations on transgenic animals, methods for the transfer of chemoresistance genes to hematopoietic cells of normal animals or patients, respectively, were developed. It has been shown that transfer the multidrug resistance gene to normal bone marrow cells with the use of retroviral vectors protects them from the toxicity of multiple drugs including taxol, colchicine, and daunomycin [32]. Similar studies were performed with the use of a retroviral vector which contained the *MRP* transporter sequence [12]. In animal models, *MDR*1-transduced murine bone marrow cells were transplanted into W/W^v or lethally irradiated normal mice [38, 47]. Both investigators detected elevated levels of *MDR*1 expression after treatment of recipient mice with taxol. This was confirmed by Hanania et al. [24] who first transplanted *MDR*1-transduced bone marrow into recipient mice, followed by taxol treatment of the recipient animals. Their bone marrow was then transplanted into a second generation of recipient mice. In several cycles of re-transplantation and taxol treatment of recipient mice, increasingly high levels of drug resistance were generated *in vivo*. Mice of the fifth and sixth generation survived extremely high doses of taxol which were lethal for normal, untreated mice. These experiments showed that highly resistant bone marrow cells were selected by drug treatment, resulting in very efficient protection from anticancer drug toxicity.

Transfer of drug resistance genes bears a risk if cancer cells contaminate bone marrow. In advanced prostate cancer, bone metastases are frequent. The risk of unwanted *MDR*1 gene transfer to cancer cells should be greatly reduced if hematopoietic stem cells are separated. Thereby, differentiated hematopoietic cells as well as non-hematopoietic cells are depleted from the cells which are capable of reconstituting bone marrow after transplantation. In addition, separation of stem cells increases the multiplicity of infection (MOI), i. e., the ratio of virus particles to target cells because short-lived hematopoietic cells that fail to repopulate bone marrow are removed. Only stem cells have the potential of unlimited life-span because they have the capacity of self-renewal, whereas more mature progenitor cells become more differentiated with each step of cell duplication and eventually undergo terminal differentiation and apoptosis.

We have transferred an *MDR*1 cDNA to a small population of mouse bone marrow cells that expressed neither lineage-specific antigens nor the MHC class II-associated I-a antigen, but had high levels of Sca-1, also referred to as Ly6A/E [30]. This cell population was separated by Ficoll density gradient centrifugation and a two-step immunomagnetic separation procedure. The stem cell phenotype of the isolated cell population was analyzed by flow cytometry, and biological stem cell properties were confirmed *in vivo* in our investigation.

The isolated cells were grown ex vivo with the use of growth factors while gene transfer was performed by coculturing with retrovirus producing fibroblasts. Functional human P-glycoprotein was detected in approximately 60% of expanded cells. After transplantation of the ex vivo-transduced cells into recipient SCID-mice, P-glycoprotein was expressed in a high proportion of bone marrow cells of the recipient animals. *MDR*1 gene expression levels were similar to those observed in multidrug-resistant cancers in the clinic. These investigations revealed that, in principle, retroviral gene transfer of chemoresistance to isolated stem cells is feasible. Comparable studies were reported by Bodine et al. [5] who transferred the *MDR*1 gene to stem cells of mice mobilized into peripheral blood by administration of growth factors.

Several clinical trials are currently being conducted on retroviral *MDR*1 gene transfer to hematopoietic cells of cancer patients [11, 26, 37]. These phase I-studies were aimed at proving efficiency, feasibility and safety of *MDR*1 gene transfer. In some trials patients were treated with taxol following reinfusion of transduced cells. However, technical limitations still impede the clinical application of this novel approach. The first reported results indicated poor gene transfer efficiency and low long-term gene expression, respectively [25]. Both retroviral vectors and the appropriate selection strategy have to be improved.

Selection of *MDR*1-transduced hematopoietic cells by exposure to anticancer drugs

Low or unstable expression of transferred genes is a major cause for failure of somatic gene therapy. Enrichment of transferred genes by drug selection should theoretically improve the efficiency of this method. To optimize the effects of drug selection on hematopoietic progenitor cells, we have investigated the feasibility of short-term selection *in vitro* following retroviral transfer of the *MDR*1 gene to bone marrow cells. It was found that short periods of drug treatment (48h to 96h) increased the expression of *MDR*1 mRNA and of P-glycoprotein in cultured bone marrow cells. As shown by Southern blot analysis, this increase was not the result of increased transcription but rather due to clonal selection of the *MDR*1 transduced hematopoietic cells. Correspondingly, drug resistance of drug-selected hematopoietic cells was increased as compared to cells that were transduced under identical conditions without subsequent drug selection [31].

Increase of gene expression by drug selection may be helpful not only in gene therapy of cancer but also of genetically determined diseases. To this end, vectors have been engineered in which the multidrug resistance gene is coexpressed with other genes. With the use of an internal ribosomal entry-site (IRES) originating from encephalomyocarditisvirus, two cDNAs can be translated from a single mRNA transcript into two separated proteins (Sugimoto et al., 1994). Such bicistronic transcriptional fusion vectors were constructed containing *MDR*1 and either the glucocerebrosidase gene or the α-galactosidase gene, which are defective in Gaucher or Fabry disease, respectively. Enrichment of transduced cells was facilitated by cytotoxic drug selection [1, 50]. With a suitable selection strategy, complete restoration of the underlying enzyme deficiency could be accomplished in cells of Gaucher patients [3].

For further characterization of such vectors, a gene-marking vector has been constructed in which *MDR*1 is combined with the lacZ gene encoding an enzyme, β-galactosidase [2]. Transduced cells are readily detectable with a histochemical reaction. This vector may be helpful to optimize selection strategies *in vivo*.

Conclusion

Multiple causes may contribute to anticancer drug resistance. Strategies to overcome chemoresisistancer are aimed at interfering with transcription or function of chemoresistance genes. While strategies developed to pharmacologically inhibit drug transport by P-glyco-protein have already been introduced into clinical treatment, molecular approaches are still at the stage of preclinical models.

Different from attempts to overcome drug resistance, more recently research has also focused on potential applications for chemoresistance genes in gene therapy. Transfer of drug resistance genes may protect hematopoietic cells from the toxicity of anticancer drugs. In addition, they may be useful tools to introduce and overexpress otherwise non-selectable genes into bone marrow. Thereby unstable expression of therapeutic genes which currently impedes efficient gene therapy of inherited disorders may be improved.

Acknowledgement

Critical reading of the manuscript by Prof. Dr. L. Bergmann is appreciated.

References

[1] Aran, J. M., M. M. Gottesman, I. Pastan: Drug-selected coexpression of human glucocerebrosidase and P-glycoprotein using a bicistronic vector. Proc. Natl. Acad. Sci. USA *91* (1994) 3176–3180.
[2] Aran, J. M., T. Licht, M. M. Gottesman, et al.: Design of a multidrug-resistance – marking vector. Proc. Am. Assoc. Cancer Res. *36* (1995) 336.
[3] Aran, J. M., T. Licht, M. M. Gottesman, et al.: Complete restoration of glucocerebrosidase deficiency in Gaucher fibroblasts using a bicistronic MDR retrovirus and a new selection strategy. Hum. Gene Ther. *7* (1996) 2165–2175.
[4] Bashir, I., K. Sikora, P. Abel, et al.: Establishment and in vivo characterization of multidrug-resistant Dunning R3327 rat prostate-carcinoma cell-lines. Int. J. Cancer *57* (1994) 719–26.
[5] Bodine, D. M., N. E. Seidel, M. S. Gale et al.: Efficient retrovirus transduction of mouse pluripotent hematopoietic stem cells mobilized into the peripheral blood by treatment with granulocyte colony-stimulating factor and stem cell factor. Blood *84* (1994) 1482–1491.
[6] Chaudhary, P. M., I. B. Roninson: Induction of multidrug resistance in human cells by transient exposure to different chemotherapeutic drugs. J. Natl. Cancer Inst. *85* (1993) 632–639.
[7] Chin K.-V., S. Chauhan, I. Pastan et al.: Regulation of *mdr* RNA levels in response to cytotoxic drugs in rodent cells. Cell Growth Differ *1* (1990) 361–365.
[8] Cole, S. P. C., G. Bhardwaj, J. H. Gerlach et al.: Overexpression of a transporter gene in a multidrug-resistant human lung cancer cell line. Science *258* (1992) 1650–1654.
[9] Cole, S. P., K. E. Sparks, K. Fraser et al.: Pharmacological characterization of multidrug resistant MRP-transfected human tumor cells. Cancer Res. *54* (1994) 5902–5910.
[10] Corrias, M. V., G. P. Tonini: An oligomer complementary to the 5' end region of the MDR1 gene decreases resistance to doxorubicin of human adenocarcinoma multidrug-resitant cells. Anticancer Res. *12* (1992) 1431–1438.
[11] Deisseroth, A. B., J. Kavanagh, R. Champlin: Use of safety-modified retroviruses to introduce chemotherapy resistance sequences into normal hematopoietic cells for chemoprotection during the therapy of ovarian cancer: a pilot trial. Hum. Gene Ther. *5* (1994) 1507–1522.

[12] D'Hondt V., M. Caruso, M. Ward et al. Expression of the multidrug resistance gene using retroviral vectors. Blood *86* (1995) 466A.
[13] Dicker, A. P., M. Volkenandt, B. I. Schweitzer et al.: Identification and characterization of a mutation in the dihydrofolate reductase gene from the methotrexate resistant CHO cell line Pro^{-3} MTX^{RIII}. J. Biol. Chem. *265* (1990) 8317–8321.
[14 Efferth, T., M. Volm: Modulation of P-glycoprotein mediated multidrug resistance by monoclonal antibodies, immunotoxins or antisense oligodeoxynucleotides in kidney carcinoma and normal kidney cells. Oncology *50* (1993) 303–308.
[15] Endicott, J. A., V. Ling: The biochemistry of P-glycoprotein-mediated multidrug resistance. Annu. Rev. Biochem. *58* (1989) 137–171.
[16] Fleming, G. F., J. M. Amato, M. Agresti et al. Megestrol acetate reverses multidrug resistance and interacts with P-glycoprotein. Cancer Chemother. Pharmacol. *29* (1992) 445–449.
[17] Ford, J. M., W. C. Prozialeck, W. N. Hait: Structural features determining activity of phenothiazines and related drugs for inhibition of cell growth and reversal of multidrug resistance in human and murine cell lines. Mol. Pharmacol. *35* (1989) 105–115.
[18] Ford, J. M., E. P. Bruggemann, I. Pastan et al.: Cellular and biochemical characterization of thioxanthenes for reversal of multidrug resistance in human and murine cell lines. Cancer Res. *50* (1990) 1748–1756.
[19] Galski, H., M. Sullivan, M. C. Willingham et al.: Expression of a human multidrug resistance cDNA (MDR1) in the bone marrow of transgenic mice: resistance to daunomycin induced leukopenia. Mol. Cell Biol. *9* (1989) 4357–4363.
[20] Gosland, M. P., B. L. Lum, B. I. Sikic: Reversal by cefoperazone of resistance to etoposide, doxorubicin and vinblastine in multidrug resistant human sarcoma cells. Cancer Res. *49* (1989) 6901–6905.
[21] Gottesman, M. M., I. Pastan: Biochemistry of multidrug resistance mediated by the multidrug transporter. Annu. Rev. Biochem. *62* (1993) 385–427.
[22] Grogan, T. M., C. M. Spier, S. E. Salmon et al.: P-glycoprotein expression in human plasma cell myeloma: correlation with prior chemotherapy. Blood *81* (1993) 490–495.
[23] Haber, D. A., S. M. Beverly, M. L. Kiely et al.: Properties of an altered dihydrofolate reductase encoded by amplified genes in cultured mouse fibroblasts. J. Biol. Chem. *256* (1981) 9501–9510.
[24] Hanania, E. G., A. B. Deisseroth: Serial transplantation shows that early hematopoietic precursor cells are transduced by MDR-1 retroviral vector in a mouse gene therapy model. Cancer Gene Therapy *1* (1994) 21–25.
[25] Hanania, E. G., R. E. Giles, J. Kavanagh et al.: Results of MDR-1 vector modification trial indicate that granulocyte-macrophage colony-forming unit cells do not contribute to posttransplant hematopoietic recovery following intensive chemotherapy. Proc. Natl. Acad. Sci. USA *93* (1996) 15346–15351.
[26] Hesdorffer, C., K. Antman, A. Bank et al.: Human MDR1 gene transfer in patients with advanced cancer. Hum. Gene Ther. *5* (1994) 1151–1160.
[27] Kobayashi, H., T. Dorai, J. F. Holland, T. Ohnuma: Cleavage of human MDR1 by a hammerhead ribozyme. FEBS Lett. *319* (1993) 71–74.
[28] Kurosawa, M., M. Okabe, N. Hara et al.: Reversal effect of itraconazole on adriamycin and etoposide resistance in human leukemia cells. Ann. Hematol. 72 (1996) 17–21.
[29] Licht, T., H. H. Fiebig, K. J. Bross et al.: Induction of multiple-drug resistance during anti-neoplastic chemotherapy in vitro. Int. J. Cancer *49* (1991) 630–637.
[30] Licht, T., I. Aksentijevich, M. M. Gottesman et al.: Efficient expression of functional human *MDR*1 gene in murine bone marrow after retroviral transduction of purified hematopoietic stem cells. Blood *86* (1995) 111–121.

[31] Licht, T., J. M. Aran, S. K. Goldenberg et al.: The MDR1 (multidrug resistance) gene acts as a drug-selectable marker in hematopoietic progenitor cells: implications for somatic gene therapy of hematopoietic disorders. Br. J. Heaematol. *93* (suppl. 2) (1996) 336.
[32] McLachlin, J. R., M. A. Eglitis, K. Ueda et al.: Expression of a human complementary DNA for the human multidrug resistance gene in murine hematopoietic precursor cells with the use of retroviral gene transfer. J. Natl. Cancer Inst. *82* (1990) 1260–1263.
[33] Mickisch, G. H., T. Licht, G. T. Merlino et al.: Chemotherapy and chemosensitization of transgenic mice which express the human multidrug resistance gene in bone marrow: Efficacy, potency, and toxicity. Cancer Res. *51* (1991a) 5417–5424.
[34] Mickisch, G. H., G. T. Merlino, H. Galski et al.: Transgenic mice which express the human multidrug resistance gene in bone marrow enable a rapid identification of agents which reverse drug resistance. Proc. Natl. Acad. Sci. USA *88* (1991b) 547–551.
[35] Mickisch, G. H., I. Aksentijevich, P. V. Schoenlein et al.: Transplantation of bone marrow cells from transgenic mice expressing the human MDR1 gene results in long-term protection against the myelosuppressive effect of chemotherapy in mice. Blood *79* (1992) 1087–1093.
[36] Oesterling, J., Z. Fuks, C. T. Lee et al.: Cancer of the prostate. In: V. T. DeVita Jr., S. Hellman, S. Rosenberg (eds.): Cancer: principles and practice of oncology. 5th edition, pp. 1322–1386. Lippincott-Raven Publishers 1997.
[37] O'Shaughnessy, J. A., K. H. Cowan, A. W. Nienhuis et al.: Retroviral mediated transfer of the human multidrug resistance gene (MDR-1) into hematopoietic stem cells during autologous transplantation after intensive chemotherapy for metastatic breast cancer. Human Gene Ther. *5* (1994) 891–911.
[38] Podda, S., M. Ward, A. Himelstein et al.: Transfer and expression of the human multiple drug resistance gene into live mice. Proc. Natl. Acad. Sci. USA *89* (1992) 9676–9680.
[39] Ramu, A., D. Glaubiger, Z. Fuks: Reversal of acquired resistance to doxorubicin in P388 murine leukemia cells by tamoxifen and other triparanol analogues. Cancer Res *44* (1984) 4392–4395.
[40] Salmon, S. E., W. S. Dalton, T. M. Grogan et al.: Multidrug-resistant myeloma: laboratory and clinical effects of verapamil as a chemosensitizer. Blood *78* (1991) 44–50.
[41] Scanlon, K. J., H. Ishida, M. Kashani-Sabet: Ribozyme-mediated reversal of the multidrug-resistant phenotype. Proc. Natl. Acad. Sci. USA *91* (1994) 11123–11127.
[42] Schadendorf, D., A. Makki, C. Stahr et al.: Membrane transport proteins associated with drug resistance expressed in human melanoma. Am J. Pathol. *47* (1995) 1545–1552.
[43] Scheffer, G. L., P. L. Wijngaard, M. J. Flens et al.: The drug resistance-related protein LRP is the human major vault protein. Nat. Med. 1 (1995) 578–82.
[44] Siegsmund, M. J., C. Cardarelli, I. Aksentijevich et al.: Ketoconazole effectively reverses multidrug resistance in highly resistant KB cells. J. Urol. *151* (1994) 485–491.
[45] Slater, L. M., P. Sweet, M. Stupeky et al.: Cyclosporin A reverses vincristine and daunorubicin resistance in acute lymphatic leukemia ***in vitro***. J Clin Invest *77* (1986) 1405–1408.
[46] Small, E. J., S. Srinivas, B. Egan et al.: Doxorubicin and dose escalated cyclophosphamide with GM-CSF for treatment of hormone-resistant prostate cancer. J. Clin. Oncol. *14* (1996) 1617–1625.
[47] Sorrentino, B. P., S. J. Brandt, D. Bodine et al.: Selection of drug-resistant bone marrow cells in vivo after retroviral transfer of human MDR1. Science *257* (1992) 99–103.
[48] Stewart, A. J., Y. Canitrot, E. Baracchini et al.: Reduction of expression of the multidrug resistance protein (MRP) in human tumor cells by antisense phosphorothioate oligonucleotides. Biochem. Pharmacol. *51* (1996) 461–469.
[49] Sugimoto, Y., I. Aksentijevich, M. M. Gottesman et al.: Efficient expression of drug-selectable genes under control of an internal ribosome entry site. Biotechnology *12* (1994) 694–698.

[50] Sugimoto, Y., I. Aksentijevich, G. Murray et al.: Retroviral co-expression of a multidrug-resistance gene (*MDR1*) and human alpha-galactosidase A for gene therapy of Fabry disease. Hum. Gene Thera. *6* (1995) 905–915.
[51] Theyer, G., M. Schirmbock, T. Thalhammer et al.: Role of the MDR-1-encoded multiple drug resistance phenotype in prostate cancer cell lines. J. Urol. *150* (1993) 1544–1547.
[52] Tsuruo, T., K. Iida, S. Tsukagoshi et al.: Overcoming of vincristine resistance in P388 leukemia in vivo and in vitro through enhanced cytotoxicity of vincristine and vinblastine by verapamil. Cancer Res. *43* (1981) 808–813.
[53] Volm, M., M. Kastel, J. Mattern et al.: Expression of resistance factors (P-glycoprotein, glutathione S-transferase-pi, and topoisomerase II) and their interrelationship to proto-oncogene products in renal cell carcinomas. Cancer *71* (1993) 3981–3987.
[54] Yahanda, A. M., K. M. Adler, G. A. Fisher et al.: Phase I trial of etoposide with cyclosporine as a modulator of multidrug resistance. J. Clin. Oncol. *10* (1992) 1624–1634.
[55] Zaman, G. J. R., C. H. M. Versantvoort, J. J. M. Smit et al.: Analysis of the expression of MRP, the gene for a new putative transmembrane drug transporter, in human multidrug resistant lung cancer cell lines. Cancer Res. *53* (1993) 1747–1750.

PSA RT-PCR: current status and future potential

W. J. Ellis, R. L. Vessella, S. W. Melchior, P. H. Lange

Prostate carcinoma in men represents the second most common cause of cancer death in the United States [11]. If the tumor is detected early (e. g., before the tumor has had a chance to spread outside the confines of the prostate gland) the cancer can theoretically be eradicated with removal of the prostate gland. However, following radical prostatectomy in patients with clinically localized prostate cancer, approximately 30% of patients will show clinical or biochemical evidence of recurrent disease. Whether these failures are due to residual local disease or undetected distant micrometastases is unknown. Current staging for metastatic disease most commonly consists of a bone scan, which is only able to detect bone metastases of sufficient size that may disrupt a normal osteoblastic activity within the skeleton. Smaller occult metastatic lesions would go undetected.

In many tumors immunostaining methods have been applied to the detection of micro-metastases [1, 2, 9, 14]. Most commonly, these tests utilize antibodies to one or more epithelial specific antigens to detect epithelial cells in the bone marrow, blood, or lymph nodes. Furthermore, in some studies of breast cancer and neuroblastoma, identification of micrometastatic cells has prognostic value [2, 9].

The reverse transcriptase polymerase chain reaction (RT-PCR) is a sensitive technique of detecting cells expressing a particular gene. Detection of cells with this technique can be several orders of magnitude more sensitive than detection with traditional imunocytochemical techniques. The RT-PCR technique has been used to probe for occult metastases in many tumor systems including lymphoma, melanoma, and neuroblastoma [5, 8]. Exactly how to use these various assays is still a matter of considerable debate. The current status of these investigations was recently reviewed [12].

Amplification of the message for PSA allows for the detection of PSA producing cells, which are presumably of prostatic epithelial origin. The relative specificity of PSA for prostate cells brought great hope that the RT-PCR reaction could provide definitive molecular staging in patients with prostatic carcinoma. As a result, RT-PCR assays were developed by a number of different laboratories.

In our hands the RT-PCR assay appears to be an extremely sensitive and specific method for detecting prostate epithelial cells. We observed sensitivity with our assay of one prostate cancer cell per hundred million lymphocytes, or five copies of plasmid PSA cDNA. We noted a positive reaction in only two percent of our control patients without a suspicion of prostate carcinoma, suggesting the false positive rate with this assay is negligible [3].

Our data on preoperative RT-PCR testing reveals a poor correlation with pathologic stage. For example, in men with stage pT2 disease, the RT-PCR was positive in 15% of men while the test was positive in 23% of men with stage pT3 disease. There were no significant differences in the parameters of Gleason score, serum PSA, or tumor volume between the RT-PCR positive and negative patients. Interestingly, there was some variability with repetitive testing and the effect of cumulative testing appears to be significant. We observed the likelihood of obtaining a positive test result more than doubled when men were subjected to testing at multiple timepoints prior to prostatectomy. Thus in a group of patients in which four or more samples are obtained, the accumulative positive rate seems to exceed fifty percent. These findings support the hypothesis that tumor cells are intermittently present in the peripheral circulation at detectable levels in many patients.

Our follow-up is limited on patients tested with the RT-PCR assay preoperatively. Only one of the six men who was a documented treatment failure after radical prostatectomy had a positive RT-PCR test preoperatively. Thus in our hands, the RT-PCR done on peripheral blood does not appear capable of predicting ultimate clinical outcome.

In contradiction to our data, other groups, most notably the Columbia group, have shown a correlation between RT-PCR results and clinical staging in prostate cancer. For example, Olsson et al. [10] used a PSA RT-PCR assay with nested primers and showed a correlation between RT-PCR results and pathological findings in 100 patients undergoing radical prostatectomy. They found the RT-PCR assay was superior to Gleason score and serum PSA in predicting the presence of extra capsular disease with a odds ratio of 15.2, 3.2, and 5.9 respectively. However, the RT-PCR was the least predictive of treatment failure at one year with a odds ratio of 3.6, 7.2, and 13.2 respectively. Similarly, Ghossein et al. [4] found a correlation between RT-PCR results and clinical stage with a positive test in 16% of nine men with apparent organ confined disease and 30% in men with locally advanced or node positive disease.

Other investigators have found the RT-PCR less useful. Sokoloff and colleagues [13] utilized a quantitative assay and found a positive test in 59% or men with pathologically organ confined disease versus 72% with extra capsular extension, a difference which was not statistically significant. Ignatoff et al. [6] also demonstrated poor correlation between RT-PCR results and pathological staging in 48 men undergoing radical prostatectomy. The test was positive in 28% of men with positive margins and 23% of men with negative margins.

The reasons for the differing results between institutions are unclear. Different patient populations may explain some of the differences. For example, it is likely that stage migration has occurred earlier in areas of the country where PSA screening was aggressively applied. Investigators from these areas are then evaluating a population with lower volume tumors. In addition, there may be differences between the assays which play a role, as there are a number of places where these assays can diverge.

The RT-PCR assay is a great deal more complex than most assays used in clinical medicine. When attempting to detect the presence of malignant cells in the circulation, lymph nodes, or bone marrow by RT-PCR, one needs to not only select an appropriate mRNA transcript that distinguishes these cells from others in its environment, but to be cognizant of the several technical idiosyncrasies of these molecular assays. For example, it is known that cells can produce a small number of mRNA transcripts from genes not typically activated in that cell type. These are referred to as illegitimate transcripts. Accordingly, following "excessive" amplification, extracts of these cells can yield positive results. Should the frequency of positive results among controls be higher than expected, reducing the number of amplification cycles may be required for the assay to be acceptable for clinical testing. Another potential error can result when the amplification yields a product similar in size to the desired product, but of a different sequence. This is a false positive that must be recognized by an analysis other than size. Recommended approaches include: a) restricted digestion of the product into fragments of known size, b) hybridization of the product to an oligonucleotide probe where only the product with the correct sequence will hybridize and c) direct sequencing of the product. Another concern is one of contamination. Multiple negative controls are essential to assure that the end product is not due to contamination. These should consist of a water control, a control where the RT step is omitted and, where possible, a known negative specimen. Finally, one cannot over emphasize that even the slightest change in protocol can alter the limit of detection and/or the clinical usefulness of an RT-PCR assay. In our own experience, changing the thermal cycler even though staying with the same manufacturer and model number altered the limit of detection and necessitated re-optimizing the assay.

In order to assess the impact of different assays on clinical results, we have initiated a multi-institutional collaborative effort in which blinded clinical specimens are run in each of several laboratories. Very soon the results of this collaborative effort should answer the question of whether different assays account for the different results seen.

We are currently exploring the relationship between RT-PCR testing of the peripheral blood and bone marrow [7]. Iliac crest bone marrow aspirations are obtained at the time of surgery in consenting patients. Furthermore, postoperative bone marrow aspirations have been obtained on a sub-set of these patients. As with our peripheral blood specimens, control patients were all RT-PCR negative in the bone marrow. However, the rate of RT-PCR positivity increased dramatically in the bone marrow. Thus, we noted that positive RT-PCR signal in 56% of patients with organ confined tumors (pT2) and 73% of men with extra capsular extension (pT3) versus 16% and 27% respectively in the peripheral blood. In postoperative testing all patients who were RT-PCR negative in the bone marrow preoperatively remained RT-PCR negatively postoperatively. Approximately 70% of patients who were RT-PCR positive preoperatively have converted to RT-PCR negative postoperatively. A small percentage of patients remain RT-PCR positive in the bone marrow postoperatively. As a group, these patients appear to have a serious disease pathologically. We are following these patients closely to evaluate whether remaining RT-PCR positive in the bone marrow correlates with adverse clinical outcome. Wood et al. [15] have also evaluated bone marrow aspirate by PSA RT-PCR and found a positive singnal in 4 of 20 (20%) with pT2 disease and in 24 of 29 (83%) with extra prostatic disease.

Integrating the bone marrow and peripheral blood data, it appears that tumor cell dissemination is a relatively early phenomenon in prostate carcinoma. The variable nature of peripheral blood RT-PCR coupled with the fact that with accumulative testing an increasing number of patients will have at least one positive result suggests that the vascular dissemination of tumor cells may be an intermittent phenomenon. In most of these the likelihood of a positive RT-PCR test increases with increasing tumor burden, suggesting that the magnitude of tumor shedding increases with increasing tumor volume. The prostate cells appear to be sequestered in the bone marrow by an unknown mechanism. This process may be a passive process such as merely being trapped within the current eddies of the bone marrow sinusoids, or an active process possibility mediated by self surface receptors or chemotactic factors.

The bone marrow RT-PCR results seen after surgery are also intriguing. In most cases, patients who are RT-PCR positive preoperatively become RT-PCR negative postoperatively. We postulate that the cells being disseminated by many tumors are unable to persist in the bone marrow following removal of the primary tumor. On the other hand, cells from some tumors do persist and may auger a poor prognosis. Of course, the mechanisms surrounding this phenomenon remain to be elucidated.

In summary, the PSA RT-PCR assay is an extremely sensitive method of detecting disseminated prostate carcinoma cells. Our results suggest these cells are present in most patients with clinically localized prostate cancer before therapy. Thus there appears to be relatively little clinical value in obtaining these tests in men preoperatively. Whether PSA RT-PCR will be useful in the postoperative period in predicting clinical failure remains to be seen. However, there seems little doubt that using RT-PCR for PSA and then other prostate-associated markers will greatly increase our knowledge about the phenomenology of prostate cancer cell dissemination clinically. This knowledge in turn should facilitate discovery about the mechanisms involved in tumor cell bone deposition, survival, and growth.

References

[1] Bretton, P. R., M. Melamed, W. Fair et al.: Detection of occulr micrometastases in the bone marrow of patients with prostate carcinoma. Prostate *25* (1994) 108–114.

[2] Cote, R. J., P. Rosen, N. Lesser: Prediction of early relapse in patients with operable breast cancer by detection of occult bone marrow mictometastases. J. Clin. Onc. *9* (1991) 1749–1756.

[3] Ellis, W. J., R. Vessella, E. Corey et al.: Use of a reverse transcriptase polymerase chain reaction (RT-PCR) to detect prostate cancer cells in the blood. J. Urol. *155* (Suppl.) (1996) 378 A.

[4] Ghossein, R. A., H. Scher, W. Gerald et al.: Detection of circulating tumor cells in patients with localized and metastatic prostatic carcinoma: clinical implications. J. Clin. Onc. *13* (1995) 1195–1200.

[5] Hoon, D. S. B., Y. Wang, P. Dale et al.: Detection of occult melanoma cells in blood with a multiple-marker polymerase chain reaction assay. J. Clin. Onc. *13* (1995) 2109–2116.

[6] Ignatoff, J. M., M. Oefelein, B. Herz et al.: Prostate specific antigen (PSA) reverse transcriptase polymerase chain reaction (RT-PCR) assay in preoperative staging of prostate cancer. J. Urol. *155* (Suppl.) (1996) 417 A.

[7] Melchior, S. W., E. Corey, W. Ellis et al.: Early tumor cell dissemination in patients with clinically localized carcinoma of the prostate. Clin. Cancer Res. *3* (1997) 249–256.

[8] Miyajima, Y., K. Kato, S. Numata et al.: Detection of neuroblastoma cells in bone marrow and peripheral blood at diagnosis by the reverse transcriptase polymerase chain reaction for tyrosine hydroxylase mRNA. Cancer *75* (1995) 2757–2761.

[9] Moss, T. J., D. Sanders: Detection of neuroblastoma cells in blood. J. Clin. Onc. *8* (1990) 736–740.
[10] Olsson, C. A., G. DeVries, A. Raffo et al.: Preoperative reverse transcriptase polymerase chain reaction for prostate specific antigen predicts treatment failure following R. P. J. Urol. *155* (1996) 1557–1562.
[11] Parker, S. L., T. Tong, S. Bolden et al.: Cancer Statistics, 1997. CA Cancer J. Clin. *47* (1997) 5–27.
[12] Pelkey, T. J., H. Frierson, D. Bruins: Molecular and immunological detection of circulating tumor cells and micrometastases from solid tumors. Clin. Chem. *42(9)* (1996) 1369–1381.
[13] Sokoloff, M. G., C. Tso, R. Kaboo et al.: Quantitative polymerase chain reaction does not improve preoperative prostate cancer staging: a clinicopathological molecular analysis of 121 patients. J. Urol. *156* (1996) 1560–1566.
[14] Thorban, S., J. Roder, H. Nekarda et al.: Immunocytochemical detection of disseminated tumor cells in the bone marrow of patients with esophageal carcinoma. J. NCI *88* (1996) 1222–1226.
[15] Wood, D. P., Jr., E. Banks, S. Humphreys et al.: Identification of bone marrow micrometastases in patients with prostate cancer. Cancer *74* (1994) 2533–2540.

Special aspects of PSMA and PSA RT-PCR for the detection of disseminated prostate cells

B. Schmidt, M. Bendhack, R. Ackermann,
B. J. Schmitz-Dräger

Introduction

Since the introduction of reverse transcriptase-polymerase chain reaction (RT-PCR) experiments with primers specific for prostate specific antigen (PSA) [11] and the prostate specific membrane antigen (PSMA) [7] significant efforts have been made to detect disseminated prostate cancer (CaP) cells in the peripheral blood. The basic concept underlying this approach is the fact that PSA and PSMA are prostate specific. Most studies have focused on the correlation between RT-PCR and clinical data, such as capsular penetration, positive surgical margins, on the prediction of pathological stage [1, 9, 10] and the treatment failure following radical prostatectomy (RRP) [12]. In contrast, the facilities of the RT-PCR technique on the detection of micrometastases of prostate cancer in lymph nodes has been largely ignored in the literature since the first description by Degucci and coworkers [2, 3].

The prostate specific membrane antigen was isolated and cloned by Israeli and coworkers [5]. The protein product is a type II integral membrane protein, which shares 54% homology to the human transferrin receptor. The expression of this gene is strongly increased in response to hormone ablation treatment of LNCaP cells [6]. In contrast, PSA expression is strongly reduced in response to hormone ablation.

In this short communication we focus on two special aspects of RT-PCR: (1) the problem of hematogenous dissemination of prostate cancer cells during radical prostatectomy and (2) the question if RT-PCR is a suitable diagnostic technique for the detection of lymph node metastases in prostate cancer.

Material

Peripheral blood was obtained from patients one day before, thirty minutes after removal of the glands and two weeks after surgery. The patients were divided into three groups: patients with carcinoma of the prostate (CaP, n = 9), patients with benign prostate hyperplasia (BPH, n = 8), and control blood samples either from patients with metastatic prostate cancer (n = 3) or from individuals without prostatic disease (n = 5, one healthy male, two healthy females, one male patient and one female patient with metastatic renal cell carcinoma). Tumor stage and grade for the patients with prostate cancer are summarized in Table 1.

Table 1 PSMA RT-PCR results for patients with CaP according to histopathological tumor stage and tumor grade (*positive* = very strong signal)

CaP:	Patient ID	pre surgery	post surgery 1 day	post surgery 14 days	tumor stage/grade
Clinically	1	negative	negative	negative	pT2C pN0M0 G2 R0
localized CaP:	2	negative	negative	negative	pT2C pN0M0 G3 R0
	3	negative	*positive*	positive	pT3A pN0M0 G2 R1
	4	positive	positive	negative	pT4A pN0M0 G3 R1
	5	negative	negative	negative	pT2A pN2M0 G2 R0
	6	negative	positive	negative	pT2C pN2M0 G2 R1
	7	positive	*positive*	negative	pT2C pN0M0 G2 R0
	8	negative	negative	negative	pT2C pN0M0 G2 R1
	9	negative	negative	negative	pT3C pN0M0 G2 R1
met. CaP	10	positive			T4 pN2M1 G3
	11	positive			T4 NxM1 G3
	12	*positive*			T2 N3M1 G2

Lymph nodes were obtained from seven patients with localized prostate cancer and from one female patient with bladder cancer. The tissue was immediately shock frozen and 5 μm parallel sections from 30 lymph nodes were either examined by the pathologist or analysed for PSMA by RT-PCR.

Methods

Intact peripheral blood leucocytes from 2 ml venous blood were isolated after erythrocyte lysis with hypotonic buffer (erythrocyte lysis buffer ELB, Qiagen, Hilden Germany) by three centrifugation steps (10 minutes at 400 g and 4 °C). Leucocytes were either stored at −70 °C or underwent immediate RNA isolation using the RNA Blood Kit (Qiagen, Hilden, Germany). RNA was quantified by UV absorption. 2 μg total RNA were employed for cDNA synthesis using AMV (Avian Myeloblastosis Virus) reverse transcriptase (Promega, Madison, USA).

To study the intraoperative dissemination of prostate cells a semi-quantitative PCR assay for PSMA and PSA using glyceraldehyde-3-phosphate dehydrogenase (GAPDH) as internal control and as standard was performed. Primers for PSMA and GAPDH (379 bp) were used as previously published [7, 14]. To further increase the specifity of the PCR reaction in blood samples, a nested PCR assay for PSMA (234 bp) was developed. One tenth of the cDNA obtained after RT reaction was directly used for the PCR amplification.

Total RNA from lymph nodes was isolated using the TRIZOL method (Life Technologies, Eggenstein, Germany) after mechanical dissection of the lymphatic tissue and quantified as decribed above. RT-PCR for PSA (710 bp) was performed using primers described earlier [9]. Primers for GAPDH yielding a product of 620 bp were purchased from Stratagene (San Diego, CA, USA).

The PCR products were labeled during PCR cycling by incorporating 2.6 μM digoxigenine labeled dUTP and 50 μM dNTPs. After separation on a 1.6% agarose gel, the PCR products were transfered to a nylon N+ membrane (Boehringer Mannheim, Mannheim, Germany). Quantification of PSA and PSMA expression was performed using the very sensitive digoxigenine system (Boehringer Mannheim) and chemiluminescence detection.

Results

Dissemination of prostate cells by surgery

RT-PCR for PSMA from the blood of patients before transurethral resection (TUR) (n = 4) or suprapubic adenomectomy (n = 4) for BPH was positive in 1 case (12.5%) (Tab. 2). This patient remained positive after transurethral surgery. The signal was strongly increased indicating a dissemination of prostate cells during surgical treatment. The other three patients undergoing transurethral resection were negative before and after surgery. However, one patient undergoing suprapubic prostatectomy became positive after the treatment (data not shown).

Two out of nine patients (22%) with localized prostate cancer were positive before radical prostatectomy (Tab. 2). Within thirty minutes after removal of the prostate

Table 2 Detection of disseminated prostatic cells in peripheral blood of patients with BPH and CaP by PSMA RT-PCR (percent positive (positive patients/total number of patients))

		pre surgery 1 day	post surgery 1 day	post surgery 14 days
BPH:	total	12 (1/8)	25 (2/8)	
	TUR-P	25 (1/4)	25 (1/4)	n. d.
	suprapub. prostec.	00 (0/4)	25 (1/4)	n. d.
CaP	total	42 (5/12)		
	RRP	22 (2/9)	44 (4/9)	11 (1/9)
	met. CaP	100 (3/3)	n. d.	n. d.
controls:	no prostatic disease	20 (1/5)	n. d.	n. d.

the blood of 4/9 (44%) patients yielded a very strong PSMA signal, indicating significant dissemination of prostate cells into the peripheral blood. Four of these five patients became negative within 14 days after RRP. The PSMA signal of the patient remaining positive was significantly lower than immediately after surgery. This patient was pathologically classified as pT2pN0M0 G3 R0. The blood of all patients with metastatic prostate cancer was positive (Tab. 1 and 2). One male patient with metastatic renal cell carcinoma showed a strong PSMA RT-PCR signal. To further study this phenomenon another patient with metastatic renal cell carcinoma was enroled in this study. This female patient, however, was negative for PSMA.

The results of PSMA RT-PCR for one patient undergoing RRP are given in Figure 1. Amplification of the housekeeping gene GAPDH yielded a constant signal at 379 bp in all samples. Nested RT-PCR for PSMA demonstrates, that the patient was negative before RRP (lane 4). Thirty minutes after removal of the prostate, a very strong PSMA signal was detected in the peripheral blood (lane 3). 19 days after surgery only a weak signal was obtained (lane 2). The blood of a healthy male control (lane1) and the PSMA non-producing cell line PC3 (lane P) was negative. Prostate cancer cell line LnCaP, known to express PSA as well as PSMA, yielded the expected signal at 234 bp (lane L).

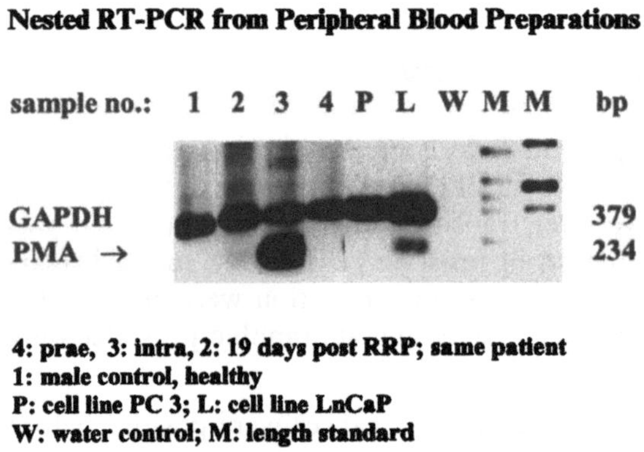

Fig. 1 Nested PSMA RT-PCR for the detection of hematogenous disseminated prostate cells during radical retropubic prostatectomy (RRP). In positive peripheral blood preparations the PSMA signal can be detected at 234 bp. GAPDH, used as an internal control, yields a constant signal at 379 bp in all samples.

Detection of micrometastases in lymph nodes

Furthermore, 30 lymph nodes from seven prostate cancer patients undergoing RRP with RT-PCR for PSA were analysed. Parallel sections of all lymph nodes were histopathologically examinated.

Histopathological examination showed micrometastases in seven lymph nodes. Using RT-PCR for PSA a signal was observed in 11 of the 30 lymph nodes (Fig. 2). While all lymph nodes of some patients were completely negative (patient no. 3), others showed only minute PSA signals (patient no. 1). Very strong signals were observed in another group of patients (patients nos. 2, 5, 6). As shown for patients 1 and 2 a strong heterogeneity between the different lymph nodes from the same patient may occur. Positive lymph nodes and negative lymph nodes may be present simultaneously.

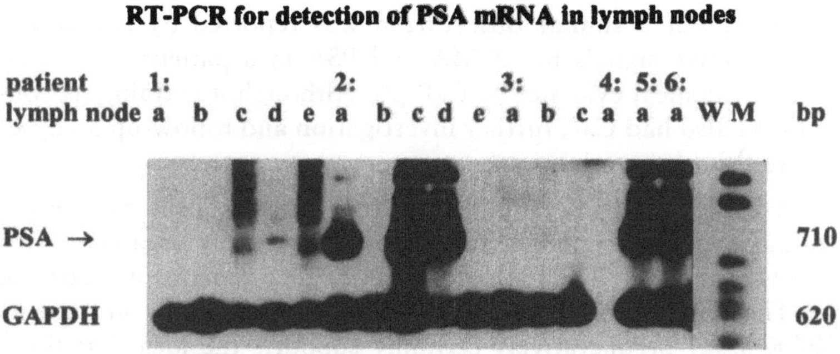

Fig. 2 PSA RT-PCR for the detection of micrometastases of prostate cancer in lymph nodes. The analysis of five lymph nodes of patients with no. 1 and 2, three lymph nodes of patient no. 3 and one lymph node each of patients with no. 4, 5 and 6 is demonstrated. In positve lymph nodes the PSA signal can be detected at 710 bp. GAPDH was used as internal control for cDNA quality (620 bp). Patient no. 4 shows no GAPDH signal, therefore, the lack of PSA signal was not interpreted as a negative result.

All pathologically positive lymph nodes were positive by RT-PCR, too. Lymph nodes from a female patient with bladder cancer undergoing lymph node dissection before radical cystectomy were PCR negative as expected (data not shown).

Conclusions

Further stimulated by the increasing incidence of prostate cancer, there is increasing clinical need to better define those patients with organ confined disease, thus allowing to specifically select those patients that may benefit from radical surgery. A variety of different parameters (serum PSA, positive core biopsies, tumor grade, etc.) were investigated for their potential prognostic value. The detection of circulating prostate cells reflecting metastatic spread appears to be a highly promising concept.

This has stimulated numerous investigators to analyse the potential of RT-PCR technology. Several tissue specific molecules, e. g. PSA, PSMA, and the recently described PTI-1 were investigated so far [13]. It has been shown, that RT-PCR is a very sensitive and specific technique for the detection of circulating prostate cells in the peripheral blood and for the detection of micrometastases in lymph nodes. However, the results obtained by different investigators vary significantly [4, 9, 12, 15]. This may be explained at least in part by the fact that all these papers differ with regard to primers and protocols, thus preventing comparison between the different studies. Therefore, standardized protocols are urgently required.

Another basic requirement is the development of internationally accepted controls. The use of standardized internal and external controls are prerequisite for developing quantitative assays. This is the basis for the introduction of "cut-off"-values which may help to eliminate false positive samples. A positive RT-PCR reaction was also seen in a patient with metastatic renal cell carcinoma inourstudy. The reasons for this finding remain unclear but a similar observation was reported by Israeli and coworkers, who found positive signals for PSMA and PSA in a patient with renal cell carcinoma but without clinical evidence of CaP [7]. Although it certainly is conceivable that these patients also had CaP, further investigation and follow-up analyses are required to delineate this observation.

Our results concerning PSMA RT-PCR demonstrate that prostate cells can be detected in the blood of patients with clinically and histopathologically organ confined CaP, but also in patients with BPH. The biological significance remains a matter of discussion until now. The observation of rising PSA levels preferentially in patients with a positive RT-PCR assay preoperatively certainly supports the idea that those cells detected by RT-PCR harbour metastatic potential [12]. Nevertheless, more data are required to finally define the clinical indication for this new assay.

Manipulation during prostate surgery for the treatment of BPH *and* CaP (TUR-P, RRP) may cause a dissemination of prostate cells into the blood. This effect was seen to a much higher degree after RRP (44%) as compared to TUR-P (25%). Approximately 80% of those patients undergoing RRP and subsequently becoming PSMA positive turned negative within 14 days after the surgery, indicating that the disseminated prostate cells were cleared from the blood. The biological significance of this observation, however, remains unclear. It may be speculated that the cell clearance from the blood rather suggests that manipulation of the prostatic gland only causes an artificial cell spill into the blood and does not result in the dissemination of potential metastatic cells as hypothesized earlier [8]. Nevertheless, the confirmation of this hypothesis certainly requires a long term follow-up.

In their first report on the use of PSA RT-PCR for the detection of micrometastatic prostate cancer cells Degucci and coworkers examined 44 lymph nodes from 22 patients (two lymph nodes/patient). RT-PCR positivity was seen in four lymph nodes, which were histologically and immunohistochemically negative [2]. Degucci and coworkers therefore concluded that the new technique could improve the diagnosis of lymph nodes metastases in prostate cancer. Our data strongly support the findings of Degucci and coworkers. However, based on our experience we recommend PSA RT-PCR analysis of more than two lymph nodes. If further follow-up analyses demonstrate that patients with PSA RT-PCR positive lymph nodes are prone to develop tumor recurrence after RRP a two stage procedure with initial (laparoscopic) lymph node dissection and PSA RT-PCR examination might help to identify proper candidates for RRP.

References

[1] Cama, C., C. A. Olsson, A. J. Raffo et al.: Molecular staging of prostate cancer II: A comparison of the application of enhanced reverse transcriptase polymerase chain reaction for the prostatic specific antigen versus prostate-specific membrane antigen. J. Urol. *153* (1995) 1373.

[2] Degucci T., T. Doi, H. Ehara et al.: Detection of micrometastatic prostate cancer cells in lymph nodes by reverse transcription-polymerase chain reaction. Cancer Res. *53* (1993) 5350.

[3] Edelstein, R. A., A. L. Zietman, R. J. Krane et al.: Implications of prostate micrometastases in pelvic lymph nodes: an archival tissue study. Urology *47* (1996) 370.

[4] Ghossein, R. A., H. I. Scher, W. K. Gerald et al.: Detection of circulating tumor cells in patients with localized and metastatic prostatic carcinoma: clinical implications. J. Clin. Oncology *13* No. 4 (1995) 1195.

[5] Israeli, R. S., C. T. Powell, W. R. Fair W. D. W. Heston: Molecular cloning of a complementary DNA encoding a prostate-specific membrane antigen. Cancer Res. *53* (1993) 227.

[6] Israeli, R. S., C. T. Powell, J. G. Corr et al.: Expression of the prostate-specific membrane antigen. Cancer Res. *54* (1993) 1807.

[7] Israeli, R. S., W. H. Miller, C. T. Powell et el.: Sensitive nested reverse transcription polymerase chain reaction of circulating prostatic tumor cells: comparisons of prostatic-specific membrane antigen and prostatic-specific antigen based assays. Cancer Res. *54* (1994) 6306.

[8] Kassabian, V. S., K. Bottles, R. Weaver et al.: Possible mechanism for seeding of tumor cells during radical prostatectomy. J. Urol. *150* (1993) 1169.

[9] Katz, A. E., C. A. Olsson, A. J. Raffo et al.: Molecular staging of prostate cancer with the use of enhanced reverse transcription PCR assay. Urology *43* (1994) 765.

[10] Katz, A. E., G. M. de Vries, M. D. Begg et al.: Enhanced reverse transcriptase-polymerase chain reaction for the prostatic-specific antigen as an indicator of true pathologic stage in patients with prostate cancer. Cancer *75* (1995) 1642.

[11] Moreno, J. G., C. M. Croce, R. Fischer et al.: Detection of hematogenous micrometastasis in patients with prostate cancer. Cancer Res. *52* (1992) 6110.

[12] Olsson, C. A., G. M. de Vries, A. J. Raffo et al.: Preoperative reverse transcription polymerase chain reaction for prostatic-specific antigen predicts treatment failure following radical prostatectomy. J. Urol. *155* (1996) 1557.

[13] Pilarski, C. P., A. E. Richter, S. Ernst et al.: Use of the detection of hematogenous micrometastases by RT-PCR of "prostate oncogene PTI-1" in the diagnosis of prostate cancer. J. Urol. *157* No. 4 (1997) 29.

[14] Schoenfeld, A., Y. Luqmani, D. Smith et al.: Detection of breast cancer micrometastases in axillary lymph nodes by using polymerase chain reaction. Cancer Res. *54* (1994) 2286.

[15] Sokoloff, M. H., C. L. Tso, R. Kaboo et al.: Quantitative polymerase chain reaction does not improve preoperative prostate cancer staging: a clinicopathological molecular analysis of 121 patients. J. Urol. *156* (1996) 1560.

Molecular forms of PSA, PSA density, age-/race-specific reference ranges, and PSA velocity

M. C. Beduschi, R. Beduschi, J. E. Oesterling

Introduction

Since its isolation and description in seminal fluid and prostatic tissue and ultimately in human serum nearly two decades ago, prostate-specific antigen (PSA) has been under intensive basic science and clinical investigation by the urologic community. As a result of such efforts and due to its unique characteristics, PSA has established itself as the most clinically useful tumor marker for the early detection and staging of PCa as well as for monitoring response to therapy [27, 37]. Though unique, PSA has important limitations that prevent it from being the ideal tumor marker for the early detection of prostate cancer [16, 28]. This comes from the fact that PSA is not cancer specific; in another words, serum PSA levels above normal are not pathognomonic for PCa nor values within the reference range are always indicative of the absence of cancer. Nevertheless, several studies have demonstrated that PSA alone detects a greater number of malignancies than digital rectal examination (DRE) or any other isolated parameter. Most importantly, the majority of such tumors have histopathological features of clinically important, life-threatening cancer [30, 34, 38].

In order to reliably detect PCa at an early stage, a low serum PSA cutoff level is used in screening programs (4.0 ng/ml). Use of this cutoff level is associated with a considerable number of false-positive findings and false-negative findings (65% false-positive and a 20% false-negative rate), demonstrating PSA's lack of power in discriminating PCa from other benign prostate pathologies (BPH) [4, 17]. On the other hand, at least 20% of men with biopsy-proven PCa at the time of diagnosis have a serum PSA concentration within the reference range of 0.0–4.0 ng/ml [15]. Had the total serum PSA concentration been the sole diagnostic criterium used, those patients would not have their cancer detected. The obvious conclusion is that the immunotechniques currently employed in the determination of serum PSA concentration have limited ability in distinguishing early PCa from BPH. This lack of sensitivity and specificity have led researchers to investigate various methods for improving the performance of total serum PSA with respect to the above cited conditions.

Several concepts have been developed and investigated recently, with the purpose of optimizing the clinical use of PSA by improving its specificity and yet preserving its sensitivity. These new techniques include *PSA density, PSA velocity, and age-/race-specific PSA reference ranges* [5, 9, 13, 29, 31]. Moreover, investigations have shown that PSA is present in the human serum in several molecular forms and that the concentration of *free and complexed PSA (PSA bound to alpha-1 antichymotrypsin)* may vary according to specific pathologic states present in the prostate gland [24,

39]. The objective of this manuscript is to review our current understanding of the above-mentioned concepts and assess their clinical utility in 1998.

Molecular forms of PSA

PSA is a chymotrypsin-like serine protease that enhances sperm motility by liquefying the semen. Its synthesis occurs almost exclusively at the prostate gland level, so for all practical purposes PSA is prostate gland-specific, but not PCa-specific [25, 35, 42]. Once in the blood circulation, this protease's function is inhibited by its binding with alpha-1 antichymotrypsin, resulting in the formation of the PSA-ACT complex [35]. Total PSA refers to the sum of all immunodetectable species of PSA (free + complexed forms). The predominant bound form of PSA is alpha-1 antichymotrypsin complex (PSA-ACT), which accounts for approximately 90% of all complexed forms of PSA [24, 39]. Free PSA possesses at least one antigenic epitope that becomes unavailable once binding with ACT occurs. Antibodies targeted against this free epitope represent the basis for the development of immunohistochemical assays that distinguish free from total PSA. Lilja et al. and Stenman et al. observed that through the quantification of free and complexed forms of PSA, they were able to enhance the specificity of PSA for detecting PCa [24, 39]. Following several investigations, it has been well demonstrated that the proportion of free PSA in the serum is lower in patients with PCa than in men with benign diseases.

Vashi et al. conducted an investigation to define the optimal total PSA range for measuring percent free-PSA (reflex range) and generated appropriate cutpoints for percent free-PSA so as to improve both the sensitivity and specificity of the PSA test in the diagnosis of PCa for men at risk, who present to the urologist for evaluation [40]. A total of 413 men with PSA value between 2.0 and 20.0 ng/ml were selected; 225 had benign disease and 188 were noted to have clinically localized PCa as determined by a sextant prostate biopsy. Afer examining all combinations of lower and upper limits of total PSA, percent free-PSA was found to offer the greatest advantage to the PSA test when the total PSA value was between 3.0 and 10.0 ng/ml (Fig. 1). The ability of percent free-PSA to distinguish early, curable cancer from BPH over this range was superior to total PSA by more than 20%. Thus, the range of 3.0–10.0 ng/ml for total PSA appears appropriate for the clinical use of percent free-PSA. This range is called the "reflex range", indicating that for these total PSA concentrations the laboratory instrument would go into a reflex mode, measuring the free-PSA level and calculating the percent free-PSA value for that specific sample. The final result is an improved sensitivity (increased cancer detection) when the total PSA value is normal and an increased specificity (eliminating negative biopsies) when the total PSA is mildly elevated.

In order to maximize sensitivity and specificity, the patients were divided into two ranges of total PSA: 3.0–4.0 ng/ml (normal total PSA) and 4.1–10.0 ng/ml (slightly elevated total PSA). For the first range (3.0–4.0 ng/ml), the objective is to enhance cancer detection (increase sensitivity) with an acceptable increase in the number of prostate biopsies performed (Fig. 2). For this group of patients, using a 19% cutoff point for percent free-PSA would result in detection of 90% of all cancers; the biopsy rate would be 73%, and the cancer detection rate would be 44%. Thus, for every 1.7 biopsies performed, one cancer is detected. For the 4.1–10.0 ng/ml range, the

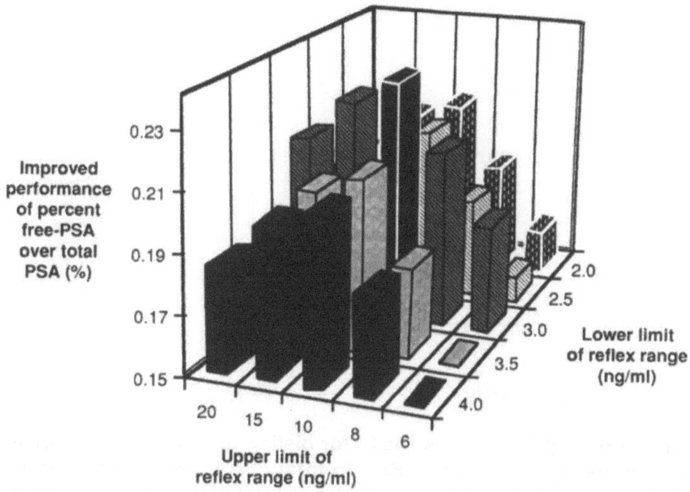

Fig. 1 Determination of the PSA "reflex range" with the best performance for percent free-PSA. Bar graph illustrating that the total PSA range of 3.0–10.0 ng/ml demonstrates the maximal improvement of percent free-PSA over total PSA. (From [40] with permission.)

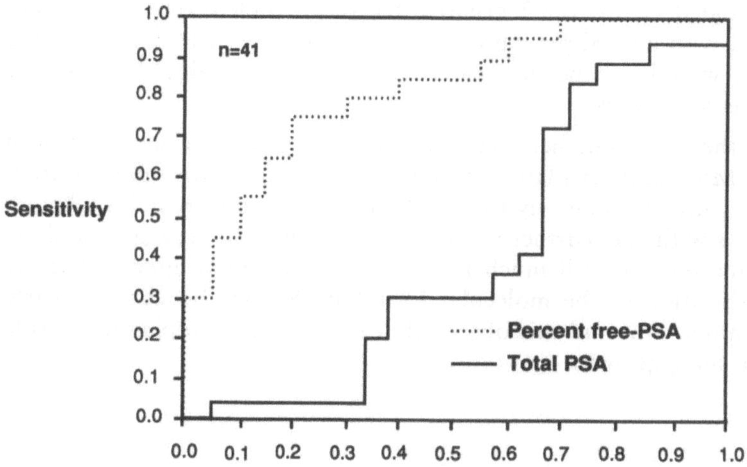

Fig. 2 Receiver operating characteristic (ROC) curves (AUC) for distinguishing prostate cancer from benign prostate disease, comparing percent free-PSA to total PSA in the "reflex range" subset of 3.0–4.0 ng/ml. The AUC curves for percent free-PSA and total PSA are 0.78 and 0.40, respectively ($p < 0.01$). (From [40] with permission.)

objective is to eliminate unnecessary prostate biopsies (increase specificity) while detecting most if not all of the PCa cases (Fig. 3).

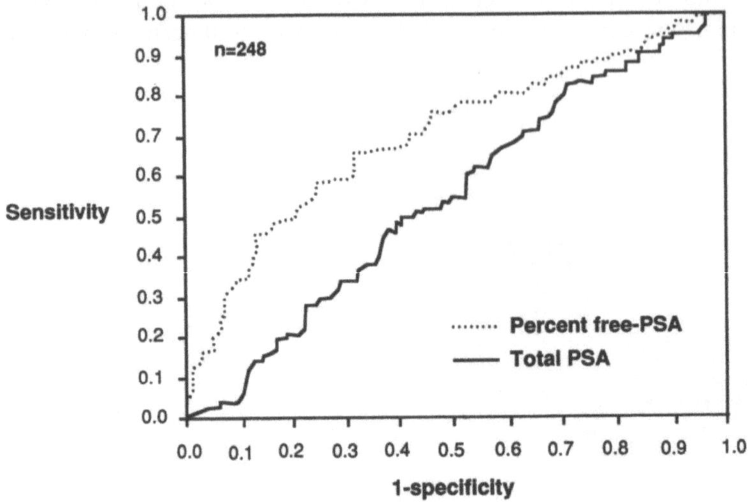

Fig. 3 Receiver operating characteristic (ROC) curves (AUC) for distinguishing prostate cancer from benign prostate disease, comparing percent free-PSA to total PSA in the "reflex range" subset of 4.1–10.0 ng/ml. The AUC curves for percent free-PSA and total PSA are 0.71 and 0.52, respectively ($p < 0.05$). (From [40] with permission.)

For this range of total PSA, the most appropriate cutpoint for percent free-PSA was 24%. At this value, 95% of the cancers would be detected and 13% of the negative biopsies could be avoided. Thus, for 7 negative biopsies avoided, one cancer goes undetected. However, with annual prostatic evaluations and use of PSA velocity, the truly significant PCa would still be detected at a curable stage (doubling time of organ-confined PCa is 3–4 years).

Concerns regarding the issue whether free PSA varies or not with age have been partially addressed. Most data available demonstrate that free PSA concentration significantly increase with age (similarly to total PSA). Controversy remains whether percent free PSA varies with age. Further investigations are now underway to address this issue. Thus, while there is still much to be learned, it would appear that the discovery and quantification of the molecular forms of PSA has brought the early detection of PCa to a new era, and this observation will continue to play a key role in PCa diagnosis and management.

PSA density (PSAD)

PSAD is mathematically defined as total serum PSA (ng/ml) divided by the volume of the prostate gland (cc^3) [7]. It is based on the assumption that PSAD calculation would theoretically adjust for the portion of PSA resulting from BPH, providing a reference value above which the increased chance of a patient having PCa would justify further investigation. This concept was first introduced in 1989 as a means of identifying preoperatively PCa patients that could not be surgically resected for cure (stage D0) [7]. Shortly thereafter, Veneziano et al. estimated a PSA-to-volume index by correlating total serum PSA and prostate volume, assessed by TRUS [41]. They reported the mean PSA-to volume index to be significantly greater in patients with PCa, indicating the usefulness of this ratio in discriminating between benign and malignant prostatic disease.

Benson et al. also reported results of 533 patients with a serum PSA level between 4.0 and 10.0 ng/ml, who were evaluated by DRE and TRUS [8]. PSAD was calculated by dividing total serum PSA by prostate volume; 289 patients underwent transrectal biopsy, of whom 98 were positive and 191 negative for PCa. The mean PSAD for the positive and negative biopsy groups was 0.297 and 0.208, respectively. An intermediate total PSA level alone was not an indication for biopsy and 45% of the patients did not undergo biopsy. Among the patients in whom biopsy was performed, there was a significant difference in the mean PSAD between those with a positive and negative biopsy. A subset of 127 patients with a normal digital rectal examination and an intermediate PSA level underwent biopsies only if transrectal ultrasound results were abnormal. In this subset, 23 patients had cancer. Mean PSADs were 0.279 and 0.204 in the cancer and benign disease subsets, respectively, and this difference was also statistically significant. It is important to note that the subgroup of patients with normal TRUS and intermediate PSA levels did not have cancer cases detected because they were not submitted to prostate biopsy.

Seaman et al. updated the initial series by Benson and analyzed 773 patients with intermediate PSA levels [36]. The overall cancer detection rate was 14.9%, but only 55% of the patients underwent biopsy. The mean PSA densities were 0.285, 0.199, and 0.165 for the positive, negative, and no biopsy groups, respectively. Again, the difference in mean PSA density between patients with a positive and negative biopsy was statistically significant, whereas the difference between mean PSA values was not. In a population without clinical evidence of PCa, the probability of a positive biopsy was 6% when the PSA density was below 0.15. On the other hand, patients with an abnormality on either DRE or TRUS, and a PSAD value equal to 0.15 the probability of a positive biopsy was 18%. The investigators concluded that the first group of patients with an intermediate PSA level, normal digital rectal examination and PSA density of less than 0.15 should be observed with PSA measurements and digital rectal examination rather than undergoing biopsy. Conversely, the group with an abnormal exam (DRE or TRUS) should always undergo a biopsy.

More recently, Bazinet et al. conducted a well designed prospective study in order to assess the role of PSAD in detecting prostate cancer [6]. Initially, 565 subjects were examined, and 142 patients were found to have a suspicious DRE and PSA levels between the 4.0 and 10.0 ng/ml range. Results after TRUS and prostate biopsy demonstrated a 16% overall rate of cancer detection. A receiver operating characteristic (ROC) curve for PSAD applied to this population revealed that the most appropriate cutoff point for indication and performance of prostate biopsies was in the range of 0.15; using this value, only 2 cancer cases would not have been identified. So, the authors concluded that PSAD can safely reduce the number of subjects submitted to systematic biopsies without significantly compromising cancer detection.

Even though much of the available data confirms its usefulness, PSAD use has failed to demonstrate improved cancer detection in some other settings, when compared to total PSA level parameters [1, 2, 10, 18]. Brawer et al., in 1992, reported his experience with 107 patients, whose PSA measurements were in the intermediate level [10]. After final analysis, PSA index did not display any significant statistical or clinical difference between the positive and negative prostate biopsy groups, with mean PSAD values of 0.192 and 0.174 for each group, respectively. They concluded that PSA index was not predictive of malignancy in these patients. Results from another study showed that 8 patients (18%) had prostate cancer, and once again no significant difference in the mean PSAD value between the patients with positive and negative biopsies was observed (mean values: 0.12 and 0.15, respectively). Also, a PSAD of 0.15 was found to be an unreliable indicator of malignancy (sensitivity: 12.5%, specificity: 61.1%, and positive predictive value: 6.7%).

In summary, the application of the PSAD concept in different patient populations for diagnostic and staging purposes resulted in enough data to demonstrate its usefulness and cost effectiveness as a *second-line* screening parameter, helping to distinguish BPH from PCa in patients with PSA levels in the intermediate range (4.0–10.0 ng/ml) and avoiding some unnecessary biopsies.

Age-/race specific reference ranges

Oesterling et al. were the first researchers to assess the relationship between age, PSA, and prostate volume. In their 1993 report, it was demonstrated that the serum PSA concentration increased over time in a population with no evidence of cancer at a ratio of 0.04 ng/ml/year [31]. In addition, they also found that serum PSA was more strongly associated with prostate volume ($r = 0.55$) than with age ($r = 0.43$). These variables placed in a multiple regression analysis showed that 30% of the serum PSA variance was due to prostate volume and 5% was accounted for by age. From these data and using the 95th percentile, age-specific reference ranges were calculated, and the recommended serum PSA ranges from this investigation, involving white man, are described in Table 1.

Table 1 Age-specific reference ranges for serum PSA

Age range (years)	Reference range (ng/ml)		
	Asians	Blacks	Whites
40–49	0.0–0.2	0.0–0.2	0.0–2.5
50–59	0.0–0.3	0.0–4.0	0.0–3.5
60–69	0.0–0.4	0.0–4.5	0.0–4.5
70–79	0.0–5.0	0.0–5.5	0.0–6.5

Oesterling postulated that using age-specific reference ranges would potentially increase the sensitivity of PSA by detecting more cases of organ-confined PCa in younger men, while increasing specificity of PSA by eliminating prostate biopsies in older men. Subsequent reports by other authors corroborated these findings, indicating that serum PSA was related to age even after adjusting for volume of the gland [20, 22].

More recently, the normal age distribution of serum PSA was evaluated by Anderson et al. in a series of 1716 men aged 40 to 79 years [3]. PSA concentrations were measured and correlated with patient age. The authors concluded that the previously published age-specific reference ranges did not adequately account for the increasing variability of PSA with age and had upper limits that were too high for men younger than 60 years of age and potentially too low for 70 to 79 year-old men. Their proposed upper limits of normality were: 1.5 ng/ml for men 40–49 years of age, 2.5 ng/ml for men 50–59 years, 4.5 ng/ml for men 60–69 years, and 7.5 ng/ml for men 70–79 years of age. These limits would still maintain 95% specificity across all ages.

In response to the age-specific reference ranges proposed by Oesterling and colleagues, Catalona et al. set out to determine whether the widely accepted cutoff point of 4.0 ng/ml needed to be adjusted for patient age [14]. A study-population of 6630 men older than 50 years of age underwent a DRE and serum PSA determination. Of this group, 1167 patients were submitted to a prostate biopsy due to either an abnormal DRE, PSA > 4.0 ng/ml, or both. Two percent (n = 264) of the men were found to have PCa, and 162 elected radical prostatectomy as the choice of treatment. Based on the pathological findings, the authors concluded that in men aged 50 to 59 years, decreasing the PSA cutoff point from 4.0 to 3.5 ng/ml would have increased the number of biopsies by 45% with a projected 15% increase in the rate of PCa detection. Moreover, in men aged 60 to 69 years, increasing the PSA cutoff point to 4.5 ng/ml would have resulted in 15% fewer biopsies at the expense of missing 8% of organ-confined tumors (2 of 25). Finally, a cutoff point of 6.5 ng/ml for men older than 70 would have translated into 44% fewer biopsies and would have missed 47% of organ-confined cancers. It was this decrease in the detection of cancer in older men that led Catalona and coworkers to conclude that the cutoff point of 4.0 ng/ml is more appropriate for older men.

Most recently, The Johns Hopkins Hospital and the Mayo Clinic joined forces to evaluate the pathologic characteristics of prostate tumors that are detected or missed using age-specific reference ranges [14]. The records of 4597 men having undergone radical prostatectomy for clinically localized PCa were reviewed. Pathologically, favorable results were characterized as organ-confined disease or capsular perforation with a Gleason score < 7, and unfavorable results presented with capsular penetration and a Gleason score > 7, or tumor involving either the seminal vesicles or lymph nodes. In the group of patients aged 60 years or less, the use of age-specific reference ranges would have detected 74 additional tumors, of which 81% were classified as favorable, whereas for the men older than 60 years, 252 fewer tumors would have been detected; 191 (76%) of them exhibited favorable pathologic results. The results of this investigation by Partin et al. supported the use of age-specific reference ranges in men less than 60 years of age.

In addition to the well-supported theory that the serum PSA concentrations correlates directly with patient age, the issue of PSA variation among different races has been investigated. As efforts continue to be put forth to optimize the ability of PSA to perform as an ideal tumor marker, investigators are working to establish age-specific reference ranges among other races.

In response to the much lower observed incidence of PCa in Asian men, Oesterling et al. set out to determine age-specific reference ranges for serum PSA from a cohort of healthy Japanese men [32]. An increase in PSA with age was observed ($r = 0.33$, $p < 0.0001$); however, the age-specific reference ranges were lower for this population. The recommended reference ranges for Japanese men are: 0–2 ng/ml for men aged 40–49 years, 0–3 ng/ml for men 50–59 years, 0–4 ng/ml for men 60–69 years, and 0–5 ng/ml for men 70–79 years of age. This difference is attributed to the fact that Japanese men have smaller prostates and the rate of prostate enlargement is less than for American men. A recent study on a large cohort of African-American men evaluated the serum PSA concentration in individuals ranging in age from 40 to 79 years [26]. The serum PSA concentration correlated directly with advancing age ($r = 0.40$, $p < 0.0001$). The suggested age-specific reference ranges for African-American to maintain 95% sensitivity at all ages are: 0.2 ng/ml in men 40–49 years, 0–4 ng/ml in men 50–59 years, 0–4.5 in men 50–59 years, 0–4.5 ng/ml in men 60–69 years, and 0–5.5 ng/ml in men 70–79 years (Table 1).

PSA velocity (PSAV)

PSAV is defined as the change in serum PSA over time. This concept was developed in an effort to improve the ability of the clinician to distinguish BPH from malignant prostatic disease, in addition to enhancing the identification rate of men with PCa destined to have progressive disease. The rationale for developing such a concept would be to use serial serum PSA measurements to evaluate men at risk of harboring PCa, thus avoiding the inter-related variables of PSA concentration, BPH, cancer volume and differentiation. It is well-known that the serum PSA concentration is directly proportional to the amount of benign epithelium and cancer volume. However, a given volume of poorly differentiated cancers produce less PSA than an equal volume of a well-differentiated prostate tumor.

In a retrospective study, Carter et al. reviewed prostatic assessment and management for 54 men participating in the Baltimore Longitudinal Study of Aging [13]. Twenty men with BPH and 18 men with PCa as determined by prostate biopsy were compared to a cohort of 16 normal individuals. The investigators calculated the PSAV for every man and found a significant difference in the age-adjusted rate of change in PSA among the 3 groups, with cancer being greater than BPH and BPH being greater than controls ($p < 0.01$) (Fig. 4). The mean change in PSA for the control group was not significantly different from zero. A PSAV $\geq$ 0.75 ng/ml per year resulted in the correct identification (sensitivity) of PCa in 72% of the individuals and accurate detection (specificity) of 92% of the BPH subjects. The authors concluded

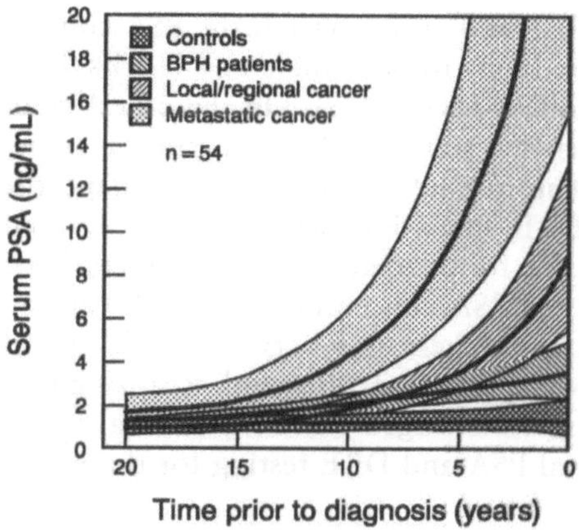

Fig. 4 Average PSA levels (95% confidence intervals) for 16 men with no evidence of prostate cancer, 20 men with a histologic diagnosis of BPH, 14 men with local/regional prostate cancer, and 4 men with metastatic prostate cancer as a function of years before diagnosis. (From [13] with permission.)

that PSAV was a potentially useful tool in distinguishing PCa from benign disease. Moreover, PSAV might be able to detect PCa in men with a normal DRE and PSA within the reference range.

Oesterling et al. conducted a prospective study where 376 men free-of-cancer had their PSA concentrations serially evaluated over a period of at least 12 months. They reported a tremendous variation in both the total PSA concentration and the PSAV across all age groups. In addition, 25–33% of men without clinical evidence of PCa had more than a 20% increase in their serum PSA levels during the study period. This finding corroborates the observations made by Carter et al. Therefore, in order to maximize the PSAV clinical value in the evaluation of men at risk of harboring PCa, it is recommended that serum PSA measurements be made at least one year apart and at least 3 determinations should be used to calculate the velocity value (Fig. 4).

In a recently published paper, Carter and colleagues set out to evaluate PSA testing intervals that would maintain the detection of curable cancer and reduce unnecessary testing [12]. In order to accomplish their objective, they conducted a historical, prospective study of serial PSA measurements at 2- and 4-year intervals from frozen serum samples of 40 men who eventually developed PCa and 272 men with no signs of prostate malignancy who were participating in an aging study. In addition, the case series of 389 consecutive men surgically treated for nonpalpable PCa was analyzed. The results reported reveal that when the serum PSA level is 4.0 ng/ml or less, nonpalpable PCa has a high probability of being cured (34/36, i. e., 94%), and the majority (25/36, 69%) were small cancers (≤ 0.5 cm^3). When the pretreatment PSA level was between 4.0 and 5.0 ng/ml, the cancers were highly likely to be curable (32/36, 89%) and a minority were small cancers (12/36, 33%). For the group of men with pretreatment PSA ≥ 5.0 ng/ml, 96 (30%) of 317 cancers were noncurable. For cancer cases, in which the baseline PSA level is less than 2.0 ng/ml, the PSA conversion to a range at which these malignancies are likely to be curable and less likely to be small (4.1 to 5.0 ng/ml) is rare at 2-year follow-up (0–4%). On the other hand, this conversion is very common when the baseline serum PSA concentration is between 2.1 and 3.0 ng/ml (27%) or 3.1 and 4.0 ng/ml (36%). Based on these data, the authors concluded that men with a normal DRE and a PSA level between 4.0 and 5.0 ng/ml, a one-year testing interval is acceptable for the detection of organ-confined (curable) PCa. When the DRE is normal and the PSA concentration is < 2.0 ng/ml, a *2-year PSA testing interval* is appropriate, since there is a very small change of missing a curable PCa when the initial PSA level is less than 2.0 ng/ml. Realizing that approximately 70% of the screened male population aged 50–70 years have PSA levels less than 2.0 ng/ml, elimination of annual PSA and DRE testing for these men would result in considerable financial savings.

Summary

In 1998, it is clear that PSA is a superior tumor marker, whether the issue is diagnosis, staging, or monitoring response to treatment. Nevertheless, it is not the ideal tumor marker. As a result, basic science and clinical investigation continue to look for new ways to enhance the clinical utility of the PSA test.

Because PSA is not specific for PCa and PCa develops in men at an age when other benign conditions are more prevalent, several concepts have been developed and are being investigated to improve the sensitivity and specificity of the PSA test. These include molecular forms of PSA (percent free-PSA), PSA density, age-specific reference ranges, and PSA velocity. All 4 parameters have their proponents and their critics and are being used at varying degrees by clinicians to improve the PCa detection rate. Regardless of the approach used, PSA testing is clearly detecting more organ-confined, clinically significant and curable PCa. Thus, given the proper time, it is reasonable to expect that the ultimate goal of any cancer detection and treatment program will be achivered – a decline of the morality rate.

References

[1] Akdas, A., T. Tarcan, L. Turkeri et al.: The diagnostic accuracy of digital rectal examination, transrectal ultrasonography, prostate specific antigen and PSA density in prostate carcinoma. Br. J. Urol. 76 (1995) 54.

[2] Akdas, A., T. Tarcan, L. Turkeri et al.: The impact of prostate-specific antigen density in predicting prostate cancer when serum prostate-specific antigen levels are less than 10 ng/ml. Eur. Urol. 29 (1996) 189.

[3] Anderson, J. R., D. Strickland, D. Corbin et al.: Age specific reference ranges for serum prostate-specific antigen. Urology 46 (1995) 54.

[4] Babain, R. J., H. A. Fritsche, R. B. Evans: Prostate specific antigen and the prostate gland volume: correlation and clinical application. J. Clin. Lab Anal. 4 (1990) 135.

[5] Babaian, R. J., H. Miyashita, R. B. Evans et al.: The distribution of prostate specific antigen in men without clinical or pathological evidence of prostate cancer: relationship to gland volume and age. J. Urol. *147* (1992) 837–840.
[6] Bazinet, M., A. A. Meshref, C. Trudel et al.: Prospective evaluation of prostate specific antigen density and systematic biopsies for early detection of prostatic carcinoma. Urology *43* (1994) 43–49.
[7] Benson, M. C., K. S. Ring, C. A. Olsson: The determination of stage D-0 carcinoma of the prostate utilizing PSA. Urologist's Correspondence Club, September 1989.
[8] Benson, M. C., I. S. Whang, C. A. Olsson et al.: The use of prostate specific antigen density (PSAD) to enhance the predictive value of intermediate levels of serum PSA. J. Urol. *147* (1992) 817.
[9] Benson, M. C., I. S. Whang, A. Pontuk et al.: Prostate specific antigen density: a means of distinguishing benign prostatic hypertrophy and prostate cancer. J. Urol. *147* (1992) 815–816.
[10] Brawer, M. K., E. A. G. Aramburu, G. L. Chen et al.: The inability of prostate specific antigen index to enhance the predictive value of prostate specific antigen in the diagnosis of prostatic carcinoma. J. Urol. *150* (1993) 369.
[11] Brawer, M. K., P. H. Lange: PSA in the screening, staging, and follow-up of early prostate cancer: a review of recent developments. World J. Urol. *7* (1989) 7.
[12] Carter, H. B., J. I. Epstein , D. W. Chan et al.: Recommended prostate-specific antigen testing intervals for the detection of curable prostate cancer. JAMA *227(18)* (1997) 1456.
[13] Carter, H. B., J. D. Pearson, E. J. Metter et al.: Longitudinal evaluation of prostate specific antigen levels in men with and without prostate disease. JAMA *267* (1992) 2215–2220.
[14] Catalona, W. J., M. A. Hudson, P. T. Scardino et al.: Selection of optimal prostate specific cutoffs for early detection of prostate cancer: Receiver-operating characteristic curves. J. Urol. *152* (1994) 2037.
[15] Catalona, W. J., D. S. Smith, T. L. Ratliff et al.: Measurement of prostate-specific antigen in serum as a screening test for prostate cancer. N. Engl. J. Med. *324* (1991) 1156–1161.
[16] Catalona, W. J., D. S. Smith, T. L. Ratliff et al.: Measurement of prostate-specific antigen in serum as a screening test for prostate cancer. N. Engl. J. Med. *324* (1991) 1156.
[17] Chan, D. W., D. J. Bruzek, J. E. Oesterling et al.: Prostate specific antigen as a marker for prostatic cancer: A monoclonal and polyclonal immunoassay compared. Clin. Chem. *33* (1987) 1916.
[18] Cookson, M. S., M. K. Floyd, T. P. Ball et al.: The lack of predictive value of prostate specific antigen density in the detection of prostate cancer in patients with normal rectal examinations and intermediate prostate specific antigen levels. J. Urol. *154* (1995) 1070.
[19] Crawford, E. D., E. P. DeAntoni: PSA as a screening test for prostate cancer. Urol. Clin. N. Amer. *20* (1993) 637.
[20] Dalkin, B. L., F. R. Ahmann, J. B. Kopp: Prostate specific antigen levels in men older than 50 years without clinical evidence of prostatic cancer. J. Urol. *150* (1993) 1837.
[21] Ercole, C. J., P. H. Lange, M. Mathisen: Prostate specific antigen and prostatic acid phosphatase in the monitoring and staging of prostate cancer. J. Urol. *138* (1987) 1181.
[22] El-Galley, R. E. S., J. A. Petros, W. A. Sanders et al.: Normal range prostate specific antigen versus age specific prostate specific antigen in screening prostate adenocarcinoma. Urology *46* (1995) 200.
[23] Hudson, M. A., R. R. Bahnson, W. J. Catalona: Clinical use of prostate specific antigen in patients with prostate cancer. J. Urol. *142* (1989) 1011.
[24] Lilja, H., A. Christensson, U. Dahlén et al.: Prostate-specific antigen in human serum occurs predominantly in complex with a-1-antichymotrypsin. Clin. Chem. *37* (1991) 1618–1625.

[25] Lundwall, A., H. Lilja: Molecular cloning of human prostate specific antigen cDNA. FEBS Letter *214* (1987) 317–322.
[26] Morgan, T. O., S. J. Jacobsen, W. F. McCarthy et al.: Age-specific reference ranges for prostate-specific antigen in African-American men. N. Engl. J. Med. *335* (1996) 304.
[27] Oesterling, J. E.: Prostate specific antigen: A critical assessment of the most useful tumor marker for adenocarcinoma of the prostate. J.Urol. *145* (1991) 907.
[28] Oesterling, J. E., D. W. Chan, J. I. Epstein: Prostate specific antigen in the preoperative and postoperative evaluation of localized prostate cancer treated with radical prostatectomy. J. Urol. *139* (1988) 766.
[29] Oesterling, J. E., W. H. Cooner, S. J. Jacobsen et al.: Influence of patient age on serum PSA concentration. Urol. Clin. N. Amer. *20* (1993) 671–680.
[30] Oesterling, J. E., W. H. Cooner, S. J. Jacobsen et al.: Influence of patient age on serum PSA concentration. Urol. Clin. N. Amer. *20* (1993) 671.
[31] Oesterling, J. E., S. J. Jacobsen, C. G. Chute et al.: Serum prostate-specific antigen in a community based population of healthy men: establishment of age-specific reference ranges. JAMA *270* (1993) 860–864.
[32] Oesterling, J. E., Y. Kumamoto, T. Tsukamoto et al.: Serum prostate-specific antigen in a community based population of healthy Japanese men: Lower values than for similarly aged white men. Br. J. Urol. *75* (1995) 347.
[33] Partin, A. W., S. R. Criley, E. N. P. Subong et al.: Standard versus age-specific prostate specific antigen reference ranges among men with clinically localized prostate cancer: A. pathological analysis. J. Urol. *155* (1996) 1336.
[34] Robles, J. M., A. R. Morell, J. P. Redorta et al.: Clinical behavior of prostatic specific antigen and prostatic acid phosphatase: A comparative study. Eur. Urol. *14* (1988) 360.
[35] Schaller, J., K. Akiyama, R. Tsuda et al.: Isolation, characterization and amino-acid sequence of gamma-seminoprotein, a glycoprotein from human seminal plasma. Eur. J. Biochem. *170* (1987) 111–118.
[36] Seaman, E., M. Whang, C. A. Olsson et al.: PSA Density (PSAD): role in patient evaluation and management. Urol. Clin. N. Amer. *20* (1993) 653.
[37] Sensabaugh, G. F.: Isolation and characterization of a semen-specific protein from human seminal plasma: A potential new marker for semen identification. J. Forensic Sci *23* (1978) 106.
[38] Sershon, P. D., M. J. Barry, J. E. Oesterling: Serum PSA values in men with histologically confirmed BPH versus patients with organ-confined prostate cancer [abstract]. J. Urol. *149* (1993) 421 A.
[39] Stenman, U. H., J. Leinonen, H. Alfthan et al.: A complex between prostate-specific antigen and alpha-1-antichymotrypsin is the major form of prostate-specific antigen in serum of patients with prostatic cancer: assay of the complex improves clinical sensitivity for cancer. Cancer Res. *51* (1991) 222–226.
[40] Vashi, A. R., K. J. Wojno, R. L. Vessella et al.: Determination of the "reflex range" and appropriate cutpoints for percent free-PA in 413 men referred for prostatic evaluation using the AxSYM system. Urology *49* (1997) 19–27.
[41] Veneziano, S., P. Pavlica, R. Querze et al.: Correlation between prostate specific antigen and prostate volume, evaluated by transrectal ultrasonography: usefulness in the diagnosis of prostate cancer. Eur. Urol. *18* (1990) 112.
[42] Wang, M. C., I. A. Valenzuela, G. P. Murphy et al.: Purification of a human prostate specific antigen. Invest. Urol. *17* (1979) 159–163.

Reverse transcriptase-polymerase chain reaction assays used for molecular staging of prostate cancer patients

C. A. Olsson, R. Buttyan

1. Introduction

The molecular technologies of polymerase chain reaction (PCR) and reverse transcriptase-PCR (RT-PCR) are incredibly powerful tools for investigative medicine. These methods are being increasingly adapted for practical use in the clinical monitoring of patients with malignancy. As early as 1988, PCR alone was utilized in the detection of residual disease in patients treated for hematologic malignancies [20, 33, 36, 62]. In these situations, tumor cells are identifiable because they have an abnormally rearranged or mutated genes that are not present in normal cells. Therefore direct PCR amplification of cellular DNA within the abnormal genetic region can identify the presence of malignant cells even when these cells are greatly diluted in a population of normal cells. The additional step of reverse-transcribing cellular messenger RNA (mRNA) to complementary DNA (cDNA) in RT-PCR enables the detection of tissue-specific cell markers in cell populations without the need for a specific mutation or gene rearrangement in the tumor cell. The utility of this technique requires the availability of cancer cell specific or tissue-specific markers.

Since prostate epithelial cells have some of the most well-characterized tissue-specific markers, prostate cancer has been the focus for the development of RT-PCR techniques that might have clinical utility. Recently, a number of different laboratories have reported on the development of RT-PCR techniques that can detect the dissemination of prostate cancer cells to metastatic sites in prostate cancer patients [11, 34, 41, 43, 57]. As early as 1990, Vessella and co-workers reported on the detection of prostate cells by RT-PCR for prostate specific antigen (PSA) in lymph nodes and bone marrow samples of patients with metastatic prostate cancer [63]. In 1992, Moreno et al. used RT-PCR to detect the presence of PSA mRNA in peripheral blood specimens of patients with metastatic prostate cancer [38]. Since then, there have been numerous reports employing this technology to detect the presence of occult prostate cells in blood, lymph nodes, and bone marrow of prostate cancer patients, and there is increasing evidence that this molecular assay can provide clinically relevant information to improve treatment for prostate cancer patients.

In an attempt to summarize the recent literature relating to these studies, this article will focus on specific areas, which include: (1) a review of RT-PCR technology, (2) what are the problems inherent to PCR-based technology, especially with regards to its use in clinical assays, (3) which prostate-specific marker are being utilized for these assays, (4) how different tissues compare for detection of prostate-specific markers, and (5) what are the areas of consensus as well as those of continued controversy in the clinical application of this assay.

2. A summary of RT-PCR technology as applied to prostate cancer testing

The purpose of the RT-PCR assays currently under development is to detect the presence of prostate cancer metastases at a far earlier time than our current diagnostic modalities allow. Although this test can be conducted upon practically any sample of body tissue, most work has been performed on the most common metastatic sites for prostate cancer, lymph nodes, and bone marrow, and on detection of occult prostate cancer cells in circulating blood. In the case of blood and marrow aspirate, the sample obtained from the patient is first processed by centrifugal separation to obtain the cell fraction containing mononuclear cells as well as any epithelial cells that might be present [7]. This cell fraction becomes the source of RNA from the specimen to which the RT-PCR procedure is applied. Although there are many techniques for RNA isolation, most used in the currently published RT-PCR studies are variations of a guanadinium-thiocyanate cell lysis procedure followed by phenol-chloroform extraction to yield total cellular RNA. New types of commercial RNA preparation products enable the rapid isolation of RNA from human tissue specimens virtually free from genomic DNA and proteins.

Once suitable RNA is obtained from a specimen, reverse transcription (RT) of the RNA to cDNA is necessary because the PCR enzyme used to amplify the target gene can only utilize a DNA template. RT is accomplished by a viral enzyme, reverse transcriptase, acting on the purified RNA in the presence of an oligonucleotide primer. There are three general types of oligonucleotides utilized for priming the RT-PCR reaction. The first, using an assortment of short random sequences to prime the RT reaction, results in the entire cellular RNA cohort (mRNAs as well as structural RNAs) being copied to cDNA. Strategies that avoid synthesizing cDNA from the structural RNAs employ either short homopolymers of dT (oligo dT) to prime cDNA synthesis exclusively from the oligo-A tails of mRNA molecules or specific antisense oligomers against the marker RNA molecule that will be the target of the RT-PCR assay. Although the latter techniques reduce the background that the target gene is diluted in, they tend to be more sensitive to partial mRNA degradation. Breaks in the RNA that separate the poly-A tail from PCR target sequences or cleave a specific priming site can prevent the synthesis of cDNA from the PCR target region. In general, the oligo-dT primer is preferred since this reagent specifically copies mRNAs and enables the subsequent PCR amplification of both the test target sequence (cell-specific marker) as well as a control target sequences (such as actin or G3PDH).

With cDNA prepared from specimen RNA, the PCR reaction can subsequently be used to amplify any target sequence present on the cDNA by repeatedly copying the DNA in a region defined by two flanking oligonucleotide primers that are included in the reaction mixture. These primers, usually 20 bases in length, are designed to hybridize to a specific site on the target cDNA sequence. The separation distance of the flanking primers on the target cDNA sequence defines the size of the amplified DNA marker. Several different computer programs are available to assist in the design and selection of PCR primers for gene sequences. For those not experienced in PCR technology, it is important to realize that primer design and selection can be the most critical aspect of a successful PCR reaction. Appropriate primer selection can also be important in distinguishing artifacts that might arise in the PCR reaction due to residual contamination of the specimen RNA with genomic DNA.

The PCR reaction is performed on a mechanical device referred to as a thermocycler because it repeatedly and reproducibly alters the temperature of the reaction tube in a manner conducive to the enzymatic amplification cycle. The reaction tubes containing single-stranded cDNA, oligonucleotide primers, a thermostable DNA polymerase, Mg^{++} ion and a nucleotide triphosphate cocktail, are warmed during an initial period so that the primers can anneal to the cDNA. Subsequently the DNA polymerase extend the primers using the cDNA as a template. An oligonucleotide annealing temperature typically in the range of 60 °C, is used to hybridize the primers to the target sequence. The reaction is then heated to 72 °C, allowing the thermostable DNA polymerases to synthesize (copy) DNA beginning from the primer oligonucleotides. Subsequently, the reaction is briefly heated to 95 °C, separating the newly synthesized DNA from the original cDNA template. The temperature is cooled to start the second cycle of amplification; in this cycle, the DNA synthesized in the first cycle becomes an additional template for the second reaction cycle. Typically, 30–40 cycles of PCR are used, each resulting in an exponential increase in target sequences present.

Variations of the PCR technique have been described that are proposed to increase the specificity of the amplification reaction. For "nested PCR" a second set of primers designed to amplify a smaller nucleotide sequence within the range defined by the initial primers can be used to re-amplify a smaller internal target sequence in a second PCR reaction. In "hot-start" PCR the polymerase is sequestered from target template until it is heated to avoid any non-specific priming and synthesis before the first PCR cycle begins.

3. Technical problems encountered with RT-PCR methods

Sample collection and handling can play critical roles in the outcome of the prostate cancer RT-PCR assays. This is because the RNA molecules that form the basis for this assay are extremely susceptible to degradation, both intracellularly and in the purified form. When cells die or lyse, nucleases that are abundantly present in clinical specimens can immediately destroy any RNA present. Therefore, appropriate preservation of specimens prior to RNA extraction is critical. This necessitates rapid processing of tissue samples from the time of collection. In our laboratory we have observed that blood samples, initially positive by RT-PCR for PSA, will convert to RT-PCR negative when stored longer than 24 hours at 4 °C. Whole blood samples appear to be more subject to storage degradation than fractionated buffy coat cells under these conditions (unpublished data). These considerations are of extreme importance for the development of a clinically suitable RT-PCR assay, particularly when dealing with a reference laboratory.

Like all PCR-based assays, RT-PCR is an extremely sensitive assay, capable of identifying the presence of one prostate cell diluted in 10^8 lymphocytes. This high degree of sensitivity mandates that extreme precaution must be taken to avoid both cross-sample and environmental contamination during specimen collection and processing. Therefore laboratory procedures that include the use of dedicated instruments and areas for performing this assay are important for preventing the occurrence of PCR-based artifacts. Other precautions include the use of laminar flow hoods for specimen processing and disposable hydrophobic-barrier pipette tips to block aerosol contamination during reaction set ups. Finally, laboratory personnel conducting the RT-PCR assays should be aware of the extreme sensitivity of the test. Obviously, a technician cannot shift from handling prostate specimens to conducting an RT-PCR directed against PSA. Even with the strictest guards against contamination, it is still necessary to use multiple controls for each RT-PCR reaction to insure the integrity of the results of this assay. Positive control specimens, often RNAs obtained from cultured prostate cancer cells (LNCaP cells), can be run in parallel with patient specimens to insure that both the RT and PCR reactions were successful. Negative controls, containing no added template cDNA, should also be run in parallel to insure against contamination of biochemical reagents and solutions, pipettes or tubes. In addition, to positive and negative controls each RNA sample must be assayed for a known housekeeping mRNA. β-actin or G3PDH primers are often used for these control reactions that ensure the integrity of the test RNA sample [30].

Because of the high sensitivity of RT-PCR, there is the distinct possibility that extremely low-level basal transcripts of prostate-specific genes might be present even in non-prostate cells and could result in a positive RT-PCR signal. Using a highly optimized 80-cycle nested RT-PCR reaction, PSA transcripts were found in blood specimens obtained from 7 normal male and 6 normal females. The same study also identified PSA transcripts, thought to be highly prostate specific, in cells lines of ovarian cancer, lung cancer, and leukemia [58]. Although PSA expression has been detected in some other forms of malignancy as well as in a normal human tissue (endometrium), the high sensitivity of the PCR reaction and its ability to detect minimal transcript levels suggest the future need to make RT-PCR assays more quantitative in order to control for basal transcription of prostate-specific markers by nonprostatic cells [24].

As was previously mentioned, one of the most crucial elements of any RT-PCR assay are the primer sequences used in the PCR reaction. Many different primer sets can often be designed that will successfully amplify and detect a given marker. However, the base-composition of the oligonucleotide primer, its location on the genomic and mRNA sequence and kinetic characteristics of a primer pair interactions will affect the sensitivity and specificity of the final PCR reaction. For example, the PSA gene, frequently used as a marker in the RT-PCR assay shares a limited sequence homology with other members of the human kallikrein gene family. The optimal primers selected for RT-PCR for PSA must be chosen from the DNA sequence in the non-homologous regions of the PSA gene in order to avoid cross-reactivity with these other mRNA species [25]. Selection of appropriate primer sequences can also be helpful in preventing artifacts due to contaminating genomic DNA in the reaction. By choosing primer pairs that span introns, an amplified genomic sequence will be larger than an amplified mRNA sequence and can be visualized as a separate band upon electrophoresis of RT-PCR products [10]. For those laboratories using RT-PCR for PSA or PSM, there is no consensus regarding a single "optimal" primer set; this single feature may help explain the discord among the results reported so far.

Likewise, there has been no unanimity regarding the assay conditions under which RT-PCR assays are conducted among the different laboratories now practicing this assay. Varying incubation times, the choice to include a RNase inhibitor and the source of RT can all effect the outcome of the RT reaction. In the PT-PCR assay, conditions such as the duration of each temperature cycle, the number of cycles, the concentration of primers, DNA polymerase and Mg^{++} ion concentration must be optimized for each set of primers.

As with specific details of the PCR method employed, techniques used for the detection of the PCR amplification products are not universally standard. Traditional techniques employ sight-based fluorescent detection of DNA bands following electrophoresis of the PCR reaction on an agarose gel using ethidium bromide-DNA intercalating dye, or Southern transfer followed by hybridization of a radiolabeled probe and autoradiography. Newer techniques including the use of digoxigenin-labeled deoxyribonucleic-triphosphates allow antibody-directed detection by chemilluminescence [30], and the C-TRAK colorimetric assay for PCR products [3] are claimed to enhance the sensitivity of PCR product detection.

Finally, the selection of patients for such assays and appropriate assignment to study categories has varied considerably in the literature. It should be recalled that PSA is an androgen-sensitive gene and that hormone therapy can result in either a decrease in mRNA transcripts or else a decrease in circulating PSA synthesizing cells [31, 74, 76]. Therefore, the use of androgen modulating therapy on a patient may alter the sensitivity of RT-PCR for PSA. When comparing investigations considering the potential that RT-PCR assays have for enhancing tumor staging, appropriate pathologic standardization is also required. A T3 lesion might be mislabeled T2 in circumstances where an exhaustive search of the linked specimen surface is not conducted [22]. Some groups have reported that various prostatic manipulations (surgery, biopsy, and digital rectal exam) can result in a positive assay for prostate-related gene products in the blood [5, 18, 39, 40, 45]. If this is correct, the timing of specimen collection compared to patient's prostate manipulation should also be coordinated and reported.

4. RT-PCR targets useful in the detection of prostate cancer

PSA remains the most noted molecule used to identify the presence of prostate cells in the RT-PCR assay. PSA is a 34,000 dalton serine protease synthesized in abundance by epithelial cells of the prostate gland and by malignant prostate cells [54, 67]. In men, PSA appears to be virtually prostate specific. In females, there have been some reports of PSA expression in the normal endometrium [6] as well as in some well documented cases of breast [9, 37] and ovarian cancer [77].

Another prostate marker that is being increasingly considered in RT-PCR assays, prostate-specific membrane antigen (PSMA), is a recently cloned and characterized 100 kd membrane bound glycoprotein [28, 42]. In contrast to PSA, which is known to be sensitive to androgen levels, PSMA expression was found to be enhanced in androgen deprived prostate cancer cells [42, 75]. This feature makes PSMA an attractive RT-PCR target in patients undergoing androgen ablation therapy, in whom PSA expression may be suppressed [42, 66]. Whereas PSA expression may be diminished or lost in highly undifferentiated tumors, PSMA expression appears to be better preserved [16].

5. Sample sources of tissue for RT-PCR assays in prostate cancer

Bone marrow and lymph nodes represent the primary metastatic sites for prostate cancer. As such, they may represent homing regions favored by prostate cancer cells that have entered either the lymphatic or venous circulation. Batson and others emphasized the importance of a pre-vertebral venous plexus affording an avenue of easy access of prostate cells to the axial skeleton [2, 12]. Warren had previously proposed that bone metastasis resulted after lymphatic spread and secondary hematogenous dissemination of prostate cancer [68]. With regards to either hypothesis, prostate cancer cells entering either lymphatic or venous circulation could theoretically be present in the peripheral blood of a patient with prostate cancer. Given the ease and simplicity of obtaining blood specimens from patients, it is no wonder that this tissue is frequently utilized for these RT-PCR assays. In fact, aside from PCR-based assays, PSA-positive cells have also been detected in bloods obtained from metastatic prostate cancer patients with the use of immunologic techniques and flow cytometry. The presence of these cells in the blood has also been correlated with the presence of metastases in patients with prostate cancer [23]. More recently, prostate cells have been harvested from the peripheral blood of patients with advanced disease and these circulating cells were confirmed as prostate cancer cells by chromosomal analysis [49].

Although these prostate cells in the circulation of prostate cancer patients appear to be cancer cells, we remain uncertain of their biological potential. Tumor metastasis is a multifactorial process. For a prostate cell to be identified in blood by an RT-PCR assay, it must have a phenotype capable of eroding through the basement membrane and migrating into the vasculature (lymphatic or venous). The cell must also be capable of surviving in the circulation, since, as previously stated, the mRNA detected by the assay would likely be rapidly degraded in a dead or dying cell [32]. Although these criteria reflect some of the necessary characteristics for tumor metastasis, other processes essential to the development of clinical metastasis are not necessarily inherent to such blood-borne prostate cancer cells, including the ability to invade distant sites, to anchor to extracellular matrix and to proliferate [1, 78]. Thus it should be emphasized that, while many prostate cancer patients have positive RT-PCR reactions, indicating the potential presence of occult prostate cancer cells in the blood, the cells detected may not reflect the entire phenotype necessary for establishing metastases. Although circulating prostate cancer cells are commonly found in patients with established metastases, when they are found in patients with clinically localized prostate cancer, they may not possess truly metastatic potential and should not yet be regarded as "micrometastases".

The frequency with which prostate cancer metastasis are found in lymph nodes and bone marrow suggests that these tissues are preferential for RT-PCR analysis as well. These sites share the necessity of requiring more invasive procedures in order to obtain appropriate samples, but several research groups have evaluated these sites as a source of material for RT-PCR analysis and have obtained intriguing evidence that such assays may be clinically valuable as well as blood-based assays. Prostate cancer cells have been putatively identified in lymph nodes by RT-PCR for PSA [8]. In this study PSA mRNA was found in five lymph nodes with identifiable metastasis and four additional nodes with negative histological and immunohistochemical analyses for prostate cells. A similar observation was made through the evaluation of archival formalin-fixed tissue wherein nodes that were thought tumor free by traditional histologic assay revealed mRNA transcripts for PSA in 44% of evaluable samples [14, 15]. RT-PCR for detection of PSA transcripts has also been used in fine needle aspiration samples of pelvic lymph nodes in patients with prostate cancer. In this study, eight of eight cytologically positive or suspicious cases were detected by RT-PCR; two of 12 cytologically negative cases also showed the presence of PSA gene products [61]. It should be emphasized that the mere detection of PSA transcripts in lymph nodes may not represent lymph node metastasis in all cases. As with any other tissue, the nodes receive both vascular and lymphatic circulation, such that circulating prostate cells are detected as they are migrating through these tissues. Thus, the suggestion that genetic analysis of lymph nodes is superior to histologic analysis remains to be proven and further clinical followup of the patient groups that were analyzed will be important to our understanding of the meaning of these results.

Wood and his colleagues have published a series of reports analyzing aspirates of bone marrow by RT-PCR for PSA in patients both with and without documented osseous metastasis [70, 71, 72, 73]. In their hands, RT-PCR assay of bone marrow provided superior information to that obtained with the use of peripheral blood for this assay. In 15 patients with untreated metastatic disease, peripheral blood was positive in 53% of cases compared to 82% of marrow samples in the same patient population. A similar observation has been made by Melchior and co-workers at the University of Washington in Seattle [35]. In their series of metastatic prostate cancer patients, 50% were RT-PCR for PSA positive by peripheral blood while 90% were positive by bone marrow. Here again, there is concern that marrow sampling represents, at least in part, sampling of the peripheral blood, since the marrow is highly vascular. Nevertheless, both of these institutions by conducting parallel assays on blood and marrow, seem to indicate that marrow sampling may be more sensitive for detecting prostate cancer metastasis than blood-based RT-PCR assays.

6. The results of clinical studies using RT-PCR assays in prostate cancer

This section reports the most recent data presented by various institutions studying RT-PCR technology, regardless of the prostate-related gene product representing the target for the technology and regardless of their focus on blood, lymph node, or marrow as the favored sample substrate. The reason for this liberal co-mingling of data is obvious. Institutions that have focused on one gene product or on one sample substrate have generally spent considerable effort to insure that the selected target or sample will "work" in their hands. When reflecting back to prior comments regarding technical difficulties with generic RT-PCR technology, this venture to co-mingle data is appropriate.

6.1 Control studies

The very first requirement for a disease testing modality that has clinical utility is that it will not label patients without the disease as "positive". Therefore appropriate analysis of controls, including patients with non-malignant prostatic disease such as benign prostatic hypertrophy, is essential to any test that is suitable for use in prostate cancer. Table 1 lists the published results of control assays in 12 different laboratories, utilizing the institutions "favored" technique (be it blood, marrow or node sampling, and using either PSM or PSA). Most groups report a 0% positivity for donor volunteers or patients with non-malignant disease of the prostate. Overall, 20 of 494 controls (4%) were positive by RT-PCR assays. While this is a low level of false positives, it should be emphasized that a single article (O'Hara et al.) accounted for 17 of the 20 false positives that were reported. It should also be recalled that the results presented in this particular study reflected an attempt by the investigators to maximally optimize the RT-PCR technology so as to identify basal-level transcription that might be present in normal individuals. By omitting this single article from review, the overall false positive rate can be reduced to 3/397 (or 0.8%).

6.2 Metastatic patients

The second requirement for a test to be clinically effective in determining prostate cancer status, is that it should approach reasonable uniformity when applied to the individual with untreated (advanced) metastatic disease. Therefore, a survey of the RT-PCR literature relative to testing the patient with confirmed metastatic prostate cancer is appropriate. We have found published data from 12 institutions from which assays of 395 patients in this category have been tested, utilizing PSA or PSM as the target marker gene for prostate cancer cells, focusing predominantly on peripheral blood sampling (two institutions utilized marrow sampling) (Table 2). Overall, 52% of patients with documented metastatic prostate cancer tested positive by RT-PCR

Table 1 Control patients

Source	RT-PCR positive	Percent	Sample	Target
Moreno et al. 1992 [11]	0/17	0%	Blood	PSA
Israeli et al. 1994 [78]	2/38	5%	Blood	PSM
	0/38	0%	Blood	PSA
Seiden et al. 1994 [79]	0/14	0%	Blood	PSA
Wood et al. 1994 [52]	0/5	0%	Marrow	PSA
Jaakkola et al. 1995 [80]	0/18	0%	Blood	PSA
Ghossein et al. 1995 [81]	0/27	0%	Blood	PSA
Olsson et al. 1996 [69]	0/65	0%	Blood	PSA
Melchior et al. 1996 [55]	0/10	0%	Marrow	PSA
Dumas et al. 1996 [82]	0/53	0%	Blood	PSA
	0/53	0%	Blood	PSM
O'Hara et al. 1996 [15]	17/97	18%	Blood	PSA
Takahashi et al. 1996 [50]	0/40	0%	Nodes	PSA
Sokoloff et al. 1996 [83)]	1/19	5%	Blood	PSA
Total	20/494	4%		

Table 2 Metastatic patients

Author	RT-PCR+	Percent	Rx	Sample	Target
Dumas et al. [71]	28/33	85%	Not reported	Blood	PSM
				Blood	PSM
Melchior et al. [55]	6/6	100%	Untreated	Marrow	PSA
Ghossein et al. [72]	25/72	35%			
Moreno et al. [11]	4/12	33%	Untreated, Refractory	Blood	PSA
Katz et al. [44]	24/28	88%	Untreated	Blood	PSA
Israeli et al. [73]	6/24	25%	Untreated, Refractory	Blood	PSA
	18/24	67%			PSM
Ellis et al. [74]	18/50	36%	Untreated, Treated, Refractory	Blood	PSA
Seiden et al. [75]	11/35	31%	Untreated, Refractory		
Jaakola et al. [76]	9/18	50%	Treated		
Slawin et al. [77]	5/6	83%	Untreated		
Sokoloff et al. [78]	29/33	88%	Untreated		
Wood et al. [51]	23/54	43%	Not reported		
Total	206/395	52%			

assays. In some instances, there was no indication as to whether any of the patients were undergoing androgen deprivation therapy at the time. Two institutions reported more favorable data using marrow over peripheral blood for RT-PCR [35, 74]. Two others suggested that using PSM as a target improved sensitivity compared to RT-PCR for PSA [27, 28]. Confining the analysis of detection capability to studies in which the authors' "favorite" assay and sample source were utilized, and in whom treatment status was indicated, 148/231 (64%) of the metastatic patients were RT-PCR positive. Further confining the analysis to those studies which stringently evaluated only untreated metastatic patients, 64/73 (88%) were positive by RT-PCR assay [35, 32, 17, 55].

6.3 Molecular staging of localized prostate cancer patients

Correlation of RT-PCR assay with pathologic stage in clinically localized prostate cancer patients has been popularized as "molecular staging" of a solid neoplasm [30]. In 1996, there will be nearly 312 000 new cases of prostate cancer detected in the USA [48]. As a consequence, nearly 165 000 patients will be subjected to radical prostatectomy this year [46]. Overall, as many as 40% of patients will be upstaged subsequent to radical surgery, suggesting that treatments in addition to radical prostatectomy will be required to control their prostate cancer [64]. The techniques currently utilized to stage prostate cancer patients are notable for their inaccuracies (DRE, CT scan, endorectal MRI) [26, 51, 52]. RT-PCR technology, if accurate as an index of stage, could greatly benefit in the selection of patients for surgery versus alternate forms of treatment when the diagnosis of clinically localized prostate cancer has been established.

Unfortunately, widely disparate reports to efficacy of RT-PCR technology in "molecular staging" of prostate cancer have been reported. As can be seen by the results shown in Table 3, most groups report a gradual increase in RT-PCR positivity (regardless of RT-PCR target or sample source) related to increasing tumor stage. However, at many institutions, the sensitivity of RT-PCR technology is sufficiently low so as not to detect those patients who might have extracapsular disease; whereas in other institutional reports, the sensitivity of RT-PCR technology is such as to create too many false positives in those patients with organ confined disease. Table 3 presents a summary of the relationship between RT-PCR and pathologic stage of prostate cancer from published data.

It can easily be seen that there is a distinct tendency of increased assay positivity with increasing stage amongst the various studies. However, because of the considerations already mentioned, there are only a few institutions where RT-PCR could be regarded as a statistically significant modality that might be used in pre-determining pathologic stage of the disease [65, 74]. For example, in the Columbia experience, if the RT-PCR assay is negative, the patient can be assured of a favorable post-operative pathology report (organ or specimen confined disease) in 88% of cases. Furthermore at this site, there has been an effort to characterize the patient population by a variety of different variables, including pre-operative serum PSA level, Gleason score and

Table 3 RT-PCR and pathologic stage

Author	pT1-pT2		pT3+		Sample	Target
Israeli et al. [77]	0/18	(0%)	1/15	(7%)	Blood	PSA
	13/18	(72%)	9/15	(60%)	Blood	PSM
Ghossein et al. [76]	4/25	(16%)	3/10	(30%)	Blood	PSA
Seiden et al. [79]	3/41	(7%)	2/8	(25%)	Blood	PSA
Wood et al. [55]	13/44	(29%)	26/42	(62%)	Marrow	PSA
Dumas et al. [75]	12/27	(44%)			Blood	PSM
	3/27	(11%)			Blood	PSA
Melchior et al. [59]	23/43	(53%)	22/27	(81.5%)	Marrow	PSA
	6/43	(14%)	10/27	(37%)	Blood	PSA
de Vries et al. [67]	12/119	(10%)	48/81	(59%)	Blood	PSA
Sokoloff et al. [82]	32/57	(56%)	12/13	(92%)	Blood	PSA
Slawin et al. [81]	27/117	(23%)	18/59	(31%)	Blood	PSA
Overall	148/579	(26%)	151/297	(51%)		
Favorites	139/491	(28%)	140/255	(55%)		

RT-PCR status. In this institution's experience, RT-PCR for PSA performed on peripheral blood samples provides statistically unique information potentially helping the pre-operative staging of the radical prostatectomy candidate.

More recently, our group at Columbia University has attempted to enhance the predictive ability of RT-PCR assays by combining these results with other observations. For example, when the serum PSA level is above 10 ng/ml and RT-PCR is negative, 81% of cases will result in specimen or organ confined pathology. In contrast, when the serum PSA is greater than 10 ng/ml and RT-PCR is positive, the patient has a 90% chance of having positive surgical margins, seminal vesicle involvement or lymph node dissemination of cancer. Furthermore, the distinction between these two groups can be analyzed by follow-up data which shows that, in the first case, the PSA failure rate is 7% compared to 43% in the latter group [65].

Additionally, a recent review of the pathology of the patient tumors evaluated by the Columbia group has suggested that the simple combination of a positive RT-PCR result along with pre-operative needle biopsy evidence of peri-neural invasion (PNI) is highly predictive of p T3 or greater disease in patients thought to have clinically localized prostate cancer. By itself the RT-PCR assay was 91% specific and 51% sensitive in predicting extraprostatic disease (odds ratio 12.0). In the same population, the presence of PNI was 80% specific and 36% sensitive in predicting p T3 status (odds ratio = 4.4). When RT-PCR and PNI positivity co-existed, 100% of patients had extraprostatic disease and the combined odds ratio was greater than 50 [50].

The controversy regarding the usefulness of RT-PCR procedures in prostate cancer staging should become better resolved as more institutional laboratories report their experiences. Recent studies, as yet unpublished, from Dr. Ursula Steiner at Freie

Universität, Germany, and from Dr. Eric Klein and his colleagues at the Cleveland Clinic (personal communications) are also now finding strong correlation between the results of a RT-PCR assay (for PSA and PSM) on patients with pathological stage after radical prostatectomy.

6.4 RT-PCR assays and molecular prognosis for localized prostate cancer patients

Three institutions have reported on the clinical progress of radical prostatectomy patients who were pre-operatively evaluated by RT-PCR. All three have noted a correlation between pre-operative RT-PCR status and post-operative tumor-free status versus incidence of biochemical failure [65, 74, 15, 47].

Table 4 Molecular prognosis

Author	Pre-RP RT-PCR neg PSA failure rate		Pre-RP RT-PCR pos PSA failure rate		Sample	Significance
Wood et al. [58]	2/41	(5%)	8/38	(21%)	Marrow	$p < 0.0126$
de Vries et al. [67]	5/111	(4.5%)	9/49	(18.4%)	Blood	$p < 0.006$
Edelstein et al. [53]	6/20	(30%)	14/16	(87.5%)	Nodes	
Overall	13/172	(7.6%)	31/103	(30%)		

In the Columbia experience, with an 18 month mean follow-up, 5/111 (5%) patients who had negative RT-PCR for PSA assays (conducted on peripheral blood) experienced post-operative biochemical failure, compared to 9/49 (18%) who were pre-operatively RT-PCR positive. By Kaplan-Meier analysis, this data was demonstrated to be statistically significant at the $p < 0.006$ level. Similarly, Wood reported on follow-up of 79 patients follow radical prostatectomy, in whom bone marrow aspirate RT-PCR was conducted. If the marrow RT-PCR for PSA was negative pre-operatively, only 2/41 (5%) failed, whereas 8/38 (21%) of patients with PSA transcripts in the marrow experienced biochemical failure. In a multivariate analysis comparing the marrow RT-PCR test to biopsy Gleason score and serum PSA level as co-variates, only the RT-PCR test was statistically associated with disease-free survival [69]. Finally, Edelstein and colleagues observed similar data analyzing RT-PCR for PSA status of pelvic lymph nodes compared to treatment outcome. 6/20 (30%) patients whose nodes were RT-PCR negative failed biochemically within a five year follow-up period, compared to 14/16 (87.5%) patients whose lymph node evaluation was RT-PCR positive.

Finally, RT-PCR status may also have prognostic significance in patients with advanced disease. When the RT-PCR assay was positive before treatment, Benson found that post-treatment RT-PCR status correlated highly with PSA response to androgen deprivation [4]. Seven patients experienced post-hormone therapy conversion to a negative RT-PCR status; all had post-treatment PSAs that were undetectable. In three patients where the RT-PCR remained positive, post-treatment PSAs were uniformly above 4.0 ng/ml.

6.5 Effect of prostatic manipulation on RT-PCR results

The influence of prostatic manipulation on RT-PCR assays had been addressed by a number of different groups (Table 5).

Table 5 Effect of prostatic manipulation

Author			Manipulation	Sample	Target
Cama et al. [23]	0/25	(0%)	Cystoscopy	Peripheral Blood	PSA
	2/32	(6%)	TRUS/Bx	Peripheral Blood	PSA
Moreno et al. [25]	4/51	(8%)	TRUS/Bx	Peripheral Blood	PSA
	3/3	(100%)	TURP	Peripheral Blood	PSA
	1/3	(33%)	RP	Peripheral Blood	PSA
Moreno et al. [27]	5/14	(36%)	RP	Operative Field	PSA
Oefelein et al. [26]	20/22	(9%)	RP	Operative Field	PSA
	5/18	(28%)	RP	Peripheral Blood	PSA
Oefelein et al. [72]	10/21	(48%)	RP	Delayed Peripheral Blood	PSA
Eschwege et al. [73]	9/11	(82%)	RP	Peripheral Blood	PSM
	4/11	(36%)	RP	Delayed Peripheral Blood	PSM
Sokoloff et al. [74]	0/11	(0%)	Various	Peripheral Blood	PSA

Cama studied peripheral blood RT-PCR for PSA status before and after DRE or cystoscopy, finding no instance where pre-manipulation negativity converted as a consequence of the manipulation in 25 patients [5]. In contrast, 2 patients from a base of 32 (6%), converted to RT-PCR positivity following TRUS biopsy. Interestingly, both of these patients had BPH. Similarly, Moreno reported on the pre- and post-TRUS biopsy status of 51 patients. Of 50 who were RT-PCR for PSA negative prior to prostatic manipulation, 4 (8%) converted to RT-PCR positive [39].

Moreno reported that in three patients who underwent transurethral resection of prostate (TURP), all three (at least transiently) converted from pre-operative RT-PCR negative status to RT-PCR positive. The influence of prostatic manipulation during the course of radical prostatectomy has also been studied. Moreno, in a separate publication, reported that 36% of patients undergoing radical prostatectomy tested positive for prostate epithelial cells in operative field blood samples taken during surgery [40]. In another series, Oefelein analyzed surgical field blood samples during radical prostatectomy using RT-PCR for PSA, finding that 91% of operative fields were positive (20/22) [45]. In 5/18 (28%) patients, peripheral blood RT-PCR converted from negative to positive. In a follow-up report, Oefelein states that there is an increase in the numbers of positive RT-PCR assays in radical prostatectomy patients, from 25% pre-operatively to 50% at three months following surgery [44]. Similar data was reported from France, where, employing RT-PCR for PSM, 9/11 patients (82%) with pre-operative negativity hemo-converted during radical prostatectomy. 4/9 (44%) of these patients remained RT-PCR for PSM positive at postoperative day three [19]. In contrast, the UCLA group reported no evidence of hemoconversion in 11 patients undergoing a variety of prostate manipulations, including TRUS and RP [59].

7. Conclusions

RT-PCR, when targeted against a prostate specific marker, can be a highly specific and sensitive assay to detect the presence of occult prostate cancer cells at sites far distant from the primary tumor. The low false positive rate (0.8%) observed when combining most published studies and the high rate of detection of prostate cells in metastatic patients (88%) found in well-defined clinical studies address the basic requirements of clinically effective diagnostic modality.

Additionally, all reports indicate that RT-PCR positivity for either PSA or PSM increases with increased stage of prostate cancer, suggesting that RT-PCR assays may ultimately have value in staging the patient presumed to have clinically localized disease. Indeed, data from 2 institutions have demonstrated the unique contribution that RT-PCR technology might lend to this most vexing problem. With the suggestive data already published, and correlative data now in press, we remain optimistic that molecular staging will become a reality in the future. After all, the wide variability in techniques of RT-PCR (as well as differences in targets and samples assayed) utilized by the various laboratories now investigating this method could themselves be responsible for many of the differences reported. There is no question that standardization among laboratories is essential before clinical utility of molecular staging can be proven. Indeed, this endeavor represents the major aim of the RT-PCR consortium in the United States. Nevertheless, at our own Medical Center, we are struck by our findings that if a pre-operative RT-PCR for PSA is negative, the patient can be assured of a successful operative outcome in 88% of cases. We are also impressed by the data showing that, when the serum PSA level is over 10 and the RT-PCR assay is positive, 90% of patients undergoing radical surgery have adverse pathology reports (and, in fact, have already experienced operative failure in nearly half the cases).

There is increasing consensus that RT-PCR assays afford prognostic information which is also unique. Three institutions have extremely similar data when comparing operative or pre-operative RT-PCR strategies (employing all three potential sample sources) to treatment outcome. This may either suggest the validity of RT-PCR as a staging modality, or, alternatively, suggest that a truly metastatic phenotype is identified in patients with circulating or node/marrow harboring PSA-synthesizing cells. In either case, RT-PCR appears to already provide information not available prior to the use of this technology. In fact, in the only series comparing RT-PCR to other, more established, predictors (PSA, Gleason score) in a multivariate analysis, RT-PCR status provided the best prognostic information.

Prostate manipulation does, in fact, appear to affect the RT-PCR assay. While DRE and cystoscopy do not seem sufficiently invasive to release prostate cells into the circulation, a small number of individuals undergoing TRUS biopsy will experience this phenomenon. This observation provides a cautionary note relative to timing of sample collection for RT-PCR assays. Furthermore, a significant fraction of patients undergoing prostate manipulation during radical surgery may experience shedding of prostate cells into the operative field and release of some into the peripheral circulation. This has not been universally observed, and, even if true, may not share the significance of spontaneous RT-PCR positivity (prior to prostatic manipulation). In the one instance, a potentially metastatic phenotype is responsible for RT-PCR positivity, whereas in the other instance, PSA or PSM synthesizing cells may have no metastatic potential at all. Further studies on the biologic half-life of such cells are required before any clinical judgment can be made regarding this phenomenon.

In summary, there appears to be consensus that RT-PCR technology can differentiate controls versus patients with metastatic disease. There is increasing consensus relative to the concept of molecular prognosis. There is no consensus regarding molecular staging at the present time. Further studies employing more standardized laboratory and clinical conditions will be required to resolve this continuing controversy.

References

[1] Aznavoorian, S., A. N. Murphy, W. G. Stetler-Stevenson, L. A. Liotta: Molecular aspects of tumor cell invasion and metastasis. Cancer *71* (1993) 1368.

[2] Batson, O. V.: The function of the vertebral veins and their role in the spread of metastasis. Ann. Surg. *112* (1940) 138.

[3] Bichi, R., G. Martinazzo, L. G. Gomella et al.: C-TRAK detection of RT-PCR products in prostate cancer molecular staging. J. Urol. *155* (1996) 553 A.

[4] Benson, M. C., G. M. de Vries, A. E. Katz et al.: Enhanced RT-PCR for PSA offers unique prognostic information during androgen deprivation therapy. J. Urol. *155* (1996) 437 A.

[5] Cama, C. L., C. A. Olsson, R. E. Buttyan et al.: Molecular staging of prostate cancer III. Effects of digital rectal examination, cystoscopy, and needle biopsy on the enhanced reverse transcriptase-polymerase chain reaction assay. J. Urol. *155* (1996) 553 A.

[6] Clements, J. A. Mukhtar: Glandular kallikreins and prostate-specific antigen are expressed in the human endometrium. J. Clin. Endocrinol. Metab. *78* (1994) 1536.

[7] Cline, E., F. Vanky, U. Galili et al.: Separation and characteristics of tumor infiltrating lymphocytes in man. In: I. P. Witz, M. D. Hana (editors). Contemporary topics in immunobiology, tenth edition, p. 79. Plenum, New York 1980.

[8] Deguchi, T., T. Doi, H. Ehara et al.: Detection of micrometastatic prostate cancer cells in lymph nodes by reverse transcriptase-polymerase chain reaction. Cancer Res. *53* (1993) 5350.

[9] Diamandis, E. P., H. Yu, D. J. Southerland: Detection of prostate specific antigen immunoreactivity in breast tumors. Breast Cancer Res. Treat. *32* (1994) 301.

[10] Dieffenbach, C. W., T. M. Lowe, G. S. Dveksler: General concepts for PCR primer design. PCR methods appl. *3* (1993) S 30.

[11] Dotta, Y. H., P. T. Adams, W. R. Drobyski et al.: Sensitive detection of occult breast cancer by the reverse transcriptase-polymerase chain reaction. J. Clin. Onc. *12* (1994) 475.

[12] Dodds, P. R., V. J. Caride, B. Lytton: The Role of vertebral veins in the dissemination of prostate cancer. J. Urol. *126* (1981) 753.

[13] Dumas, F., P. Eschwege, P. Blanchet et al.: Enhanced detection of hematogenous circulating prostatic cells in patients with prostate adenocarcinoma using prostate specific membrane

antigen (PSM)-based nested reverse transcription polymerase chain reaction assay. J. Urol. *155* (1996) 506 A.

[14] Edelstein, R. A., R. J. Crane, R. K. Babayan et al.: A rapid and simple method for the detection of prostate-specific antigen mRNA in archival tissue specimens using a reverse transcriptase-polymerase chain reaction assay. Urology *45* (1995) 597.

[15] Edelstein, R. A., A. L. Zietman, A. de las Morenas et al.: Implication of prostatic micrometastases to the pelvic lymph nodes: First report of an archival tissue gene expression study. J. Urol. *155* (1996) 550 S.

[16] Ellis, D. W., S. Leffers, T. S. Davies et al.: Multiple immunoperoxidase markers in benign hyperplasia and adenocarcinoma of the prostate. Am. J. Clin. Pathol. *8* (1984) 279.

[17] Ellis, W. J., R. L. Vessella, E. Corey et al.: Use of a reverse transcriptase polymerase chain reaction (RT-PCR) to detect prostate cancer cells in the blood. J. Urol. *155* (1996) 378 A.

[18] Eschwege, P., F. Dumas, P. Blanchet et al.: Hematogenous dissemination of prostatic epithelial cells during radical prostatectomy. Lancet *346* (1995) 1528.

[19] Eschwege, P., F. Dumas, P. Blanchet et al.: Hematogenous prostatic epithelial cells microdissemination during radical prostatectomy. J. Urol. *155* (1996) 528 A.

[20] Fey, M. F., A. E. Kulozik, T. E. Hansen-Hagge et al.: The polymerase chain reaction: A new tool for the detection of minimal residual disease in hematologic malignancies. Eur. J. Cancer *27* (1991) 89.

[21] Ghossein, R. A., H. I. Scher, W. L. Gerald et al.: Detection of circulating tumor cells in patients with localized and metastatic prostatic carcinoma: Clinical Implications. J. Clin. Oncol. *13* (1995) 1195.

[22] Hall, G. D., C. E. Kramer, J. I. Epstein: Evaluation of radical prostatectomy specimens. A comparative analysis of sampling methods. Am. J. Surg. Path. *16* (1992) 315.

[23] Hamdy, F. C., J. Lawry, J. B. Anderson et al.: Circulating prostate specific antigen cells correlate with metastatic prostate cancer. Br. J. Urol. *69* (1992) 392.

[24] O'Hara, S. M., R. W. Veltri, P. Skipstuas et al.: Basal PSA mRNA levels detected by quantitative reverse transcriptase-polymerase chain reaction (Q-RT-PCR-PSA) in blood form subjects without prostate cancer. J. Urol. *155* (1996) 418 A.

[25] Henttu, P., P. Vihko: cDNA coding for the entire human prostate specific antigen shows high homologies to the human tissue kallikrein genes. Biochem. Biophys. Res. Comm. *160* (1989) 903.

[26] Hircak, H., G. C., Dooms, R. B. Jeffery et al.: Prostatic carcinoma: Staging and clinical assessment, CT, and MR imaging. Radiology *162* (1987) 331.

[27] Israeli, R. S., W. H. Miller, S. L. Su et al.: Sensitive nested reverse transcription polymerase chain reaction detection of circulating prostatic tumor cells: Comparison of prostate-specific membrane antigen and prostate-specific antigen-based assays. Cancer Res. *54* (1994) 6306.

[28] Israeli, R. S., C. T. Powell, W. R. Fair et al.: Molecular cloning of a complimentary DNA encoding a prostate-specific membrane antigen. Cancer Res. *53* (1993) 227.

[29] Jaakkola, S., T. Vornanen, J. Leinonen et al.: Detection of prostatic cells in peripheral blood: Correlation with serum concentrations of prostate-specific antigen. Clin. Chem. *41* (1995) 182.

[30] Katz, A. E., C. A. Olsson, A. J. Raffo et al.: Molecular staging of prostate cancer with the use of an enhanced reverse transcriptase-PCR assay. Urology *43* (1994) 765.

[31] Katz, A. E., G. M. de Vries, M. D. Begg et al.: Enhanced reverse transcriptase-polymerase chain reaction for prostate specific antigen as an indicator of true pathologic stage in patients with prostate cancer. Cancer 75 (1995) 1642.

[32] Katz, A. E., G. M. de Vries, C. A. Olsson et al.: Molecular staging of genitourinary malignancies. Urology, in press.

[33] Lee, M. S., K. S. Chang, F. Cabanillas et al.: Detection of minimal residual cells carrying the t (14:18) by DNA sequence amplification. Science *237* (1988) 175.

[34] Mattano, L. A. Jr., T. J. Moss, S. G. Emerson: Sensitive detection of rare circulating neuroblastoma cells by the reverse transcriptase-polymerase chain reaction. Cancer Res. *52* (1992) 4701.

[35] Melchior, S. W., E. Corey, W. J. Ellis et al.: Early tumor cell dissemination in patients with clinically localized carcinoma of the prostate (CAP). J. Urol. *155* (1996) 554 A.

[36] Miyomura, M., Y. Tanimoto, K. Morishima et al.: Detection of philadelphia chromosome-positive acute lymphoblastic leukemias by polymerase chain reaction: possible eradication of minimal residual disease by marrow transplantation. Blood *79* (1992) 1366.

[37] Monne, M., C. M. Croce, H. Yu et al.: Molecular characterization of prostate specific antigen messenger RNA expressed in breast tumors. Cancer Res. *54* (1994) 6344.

[38] Moreno, J. G., C. M. Croce, R. Fischer et al.: Detection of hematogenous micrometastasis in patients with prostate cancer. Cancer Res. *52* (1992) 6110.

[39] Moreno, J. G., J. P. Long, M. O'Hara et al.: Surgical prostatic instrumentation causes showering of prostate cells into the circulation as determined by RT-PCR of PSA. J. Urol. *155* (1996) 553 A.

[40] Moreno, J. G., P. J. Shenot, D. Shupp-Byrne et al.: Analysis of tumor spillage during radical prostatectomy using RT-PCR of prostate-specific antigen. Techniques in Urol. *2* (1996) 54.

[41] Moss, T. J., D. G. Sanders: Detection of neuroblastoma cells in blood. J. Clin. Oncol. *8* (1990) 736.

[42] Murphy, G., H. Raagde, G. Kenny et al.: Comparison of prostate-specific membrane antigen and prostate-specific antigen levels in prostate cancer patients. Anticancer Res. *15* (1995) 1473.

[43] Naito, H., N. Kuzumaki, J. Uchino et al.: Detection of tyrosine hydroxylase mRNA and minimal neuroblastoma cells by the reverse transcription-polymerase chain reaction. Eur. J. Cancer *27* (1991) 762.

[44] Oefelein, M. G. M., B. Herz, J. M. Ignatoff et al.: Longitudinal assessment by prostate specific antigen (PSA) reverse transcriptase-polymerase chain reaction (RT-PCR) of circulating prostate epithelial cells following radical prostatectomy (RRP). J. Urol. *155* (1996) 552 A.

[45] Oefelein, M. G., K. Kaul, B. Herz et al.: Molecular detection of prostate epithelial cells from the surgical field and peripheral circulation during radical prostatectomy. J. Urol. *155* (1996) 238.

[46] Olsson, C. A., E. T. Goluboff: Detection and treatment of prostate cancer: Perspective of the urologist. J. Urol. *152* (1994) 1695.

[47] Olsson, C. A., G. M. de Vries, A. J. Raffo et al.: Preoperative reverse transcriptase-polymerase chain reaction for prostate-specific antigen predicts treatment failure following radical prostatectomy. J. Urol. *155* (1996) 1557.

[48] Parker, S. L., T. M. Tong, S. Bolden et al.: Cancer statistics. Ca. Cancer J. Clin. *46* (1996) 5.

[49] Paul, O. P., G. Ts'o, S. Bova et al.: Harvest of circulating prostate cancer cells for customized decision making, monitoring and treatment of advanced prostate cancer. J. Urol. *155* (1996) 607 A.

[50] Rubin, M. A., M. Fang, M. D. Stifelman et al.: Enhanced reverse transcriptase-PCR for prostate specific antigen combined with needle biopsy results: A strong indicator of extraprostatic disease. Submitted for publication.

[51] Salo, J. O., L. Kivisaari, S. Rannikko et al.: Computerized tomography and transrectal ultrasound in the assessment of local extension of prostatic carcinoma before radical retropubic prostatectomy. J. Urol. *137* (1987) 435.

[52] Scaletscky, R., M. O. Koch, C. W. Eckstein et al.: Tumor volume and stage in carcinoma of the prostate detected by elevations in prostate specific antigen. J. Urol. *152* (1994) 129.
[53] Seiden, M. V., P. W. Kantoff, K. Krithivas et al.: Detection of circulating tumor cells in men with localized prostate cancer. J. Clin. Oncol. *12* (1994) 2634.
[54] Sensabaugh, G. F.: Isolation and characterization of a semen specific protein from human seminal plasma: A potential new marker for semen identification. J. Forensic Sci. *23* (1978) 106.
[55] Slawin, K. M., personal communication.
[56] Slawin, K. M., S. M. O'Hara, W. Song et al.: Comparison of results obtained using different RT-PCR-PSA assay methods on a single set of clinical specimens. J. Urol. *155* (1996) 417 A.
[57] Smith, B., P. Selby, J. Southgate et al.: Detection of melanoma cells in peripheral blood by means of reverse transcriptase and polymerase chain reaction. Lancet *338* (1991) 1227.
[58] Smith, M. R., S. Biggar, M. Hussein: Prostate-specific antigen messenger RNA is expressed in non-prostate cells: Implications for detection of micrometastases. Cancer Res. *55* (1995) 2640.
[59] Sokoloff, M. H., A. Patel, T. Cho-Lea et al.: Surgical manipulation does not result in hematogenous dissemination of prostate cancer [PC] cells. J. Urol. *155* (1996) 528 A.
[60] Sokoloff, M. H., T. Cho-Lea, R. Kaboo et al.: Quantitative polymerase chain reaction does not improve prostate cancer staging: A clinical-pathologic-molecular analysis of 121 patients. J. Urol. *155* (1996) 417 A.
[61] Takahashi, T., S. Hoshi, S. Orikasa: Genetic diagnosis of pelvic lymph node metastasis in prostate cancer. J. Urol. *155* (1996) 378 A.
[62] Tycko, B., J. D. Palmer, M. P. Link et al.: Polymerase chain reaction amplification of rearranged antigen receptor genes using junction-specific oligonucleotides: Possible application for detection of minimal residual disease in acute lymphoblastic leukemia. Cancer Cells 7 (1989) 47.
[63] Vessella, R. L., K. A. Blouke, J. E. Stray et al.: The use of the polymerase chain reaction to detect metastatic prostate cancer in lymph nodes and bone marrow. Proc. Amer. Assn. Cancer Res. *33* (1992) 2367.
[64] Voges, G. E., J. E. McNeal, E. A. Redwine et al.: Morphologic analysis of surgical margins with positive findings in prostatectomy for adenocarcinomas of the prostate. Cancer *69* (1992) 520.
[65] de Vries, G. M., C. A. Olsson, M. C. Benson et al.: Quantitative and qualitative RT-PCR for PSA: Molecular staging of prostate cancer. J. Urol. *155* (1996) 417 A.
[66] Wang, M. C., L. D. Papsidero, M. Kuriyama et al.: Prostate antigen: A marker for human prostate epithelias cells. Prostate 2 (1981) 89.
[67] Wang, M. C., L. A. Valenzuela, G. P. Murphy et al.: Purification of a human prostate specific antigen. Invest. Urol. *17* (1979) 159.
[68] Warren, S., P. N. Harris, R. C. Graves: Osseous metastasis of prostate cancer. Arc. Path. *22* (1936) 139.
[69] Wood, D. P., Personal communication.
[70] Wood, D. P.: The molecular staging of prostate cancer. Semin. Urol. *13* (1995) 96.
[71] Wood, D. P., E. R. Banks, S. Humphries et al.: Sensitivity of immunohistochemistry and polymerase chain reaction in detecting prostate cancer cells in bone marrow. J. Histochem. Cytochem. *42* (1994) 505.
[72] Wood, D. P., E. R. Banks, S. Humphries et al.: Identification of bone marrow micrometastases in patients with prostate cancer. Cancer Res. *74* (1994) 2533.
[73] Wood, D. P., M. Weinstein, S. Humphries et al.: Prognostic significance of circulating prostate cells. J. Urol. *155* (1996) 418 A.

[74] Wood, D. P., S. Humphries, V. M. Rangnekar: Effect of hormonal therapy on circulating prostate cells. J. Urol. *155* (1996) 553 A.

[75] Wright, G. L., B. M. Grob, P. F. Schellhammer et al.: Prostate specific membrane antigen (PSMA) is upregulated following hormone-deprivation therapy. J. Urol. *155* (1996) 692 A.

[76] Young, C. Y., P. E. Andrews, D. J. Tindall: Expression and androgenic regulation of human prostate-specific kallikreins. J. Androl. *16* (1995) 97.

[77] Yu, H., E. P. Diamandis, M. Lavesque et al.: Expression of the prostate specific antigen gene by a primary ovarian carcinoma. Cancer Res. *55* (1995) 1603.

[78] Zetter, B. R.: The cellular basis of site-specific metastasis. N. Engl. J. Med. *322* (1990) 605.

Concluding remarks

J. E. Altwein

At this year's American Urological Association meeting the John K. Lattimer lecture was entitled: Molecular Biology for the Clinical Urologist and the speaker was Carl A. Olson, who is one of the authors of this book. This more than emphasizes the relevance of this new scientific field for the practicing urologist. The information provided in this book may pass unnoticed if the following requirements are not met:

1. Broadening the basis to conceive the elements and essential principals of molecular biology. If efforts in this direction are omitted, the urologist will most likely refrain from utilizing molecular biological tools. In comparison to this task it was fairly easy to provide the background to understand the biochemistry of the mode of action of novel medicines in the past, e. g.: pure antiandrogens, LHRH-agonists, 5-alpha-reductase-inhibitors or phosphodiesterase-inhibitors. A useful basic text has been written by Christoph Wagener [9]. The dictionary from John Kendrew and Eleanor Laurence [4] and the chart-like cartoons provided by Transduction Laboratories (to order: Dianova, Postfach 30 12 50, 20305 Hamburg) are helpful. Furthermore, a small aid in this direction is the quotation of review papers down the line in this text, if available. An incomplete background information about the potential provided through molecular biology may lead to misunderstanding on the side of the urologist, e. g., concerning the application of gene therapy for metastatic carcinoma of the prostate or the risk reduction to develop prostate cancer if one eats regularly sulforaphane, a polyphenol present in high concentrations in cruciferous vegetables such as broccoli. Even if the urologist himself neglects these management efforts, his patient may have read about such new developments and demand information from his treating physician.

2. An important obstacle for the introduction of molecular biological tools in everyday practice is the cost. Such examinations would add a considerable chunk to the already high budget from sophisticated imaging or bacteriological studies and, in particular, long-term treatment. If one would include the molecular forms of PSA in the workup and follow-up of men with prostate cancer, the costs would be unbearable.

3. Furthermore, the urologist should be aware that the techniques provided through the inclusion of molecular biology in his daily practice are a moving target and only a few techniques are destined to stay. A good example in this regard are the high hopes raised with the introduction of immunological techniques in urology [3]. As far as prostate cancer is concerned, virtually none stood the test of time [1].

Subsequently, some short remarks are given to support the understanding how molecular biology could influence everyday urology and above all could be utilized for better patient care.

Cytogenetics of prostate cancer

Just recently prostate cancer became a target of cytogenetic examinations (see the contribution of *Sauter* in this book). Two important investigational techniques provide better access to cytogenetics: Fluorescence in situ hybridization (FISH) and comparative genomic hybridization (CGH). These two molecular cytogenetic methods allow for the examination of archival material, this has the advantage that the outcome of the disease is already known, e. g., the patient survived or he did not. Sauter systematically reviews the cytogenetic alterations in prostate cancer. He points out that the Y-loss indicates an unfavorable prognosis, however, the Y-loss is also common in BPH. Another alteration is the polysomy 7, also indicative of a pure prognosis, but this reflects an alteration of the entire genome and is not useful for prostatic biopsies. Interesting is the correlation of chromosome 7 q 31.1-deletion to node positive carcinoma of the prostate. If this correlation would be in context with other prognostic parameters such as Gleason-score, it would help to spare some patients a pelvic lymphadenectomy.

Mapping studies have located positive tumor suppressor gens on chromosome 8p. However, chromosome 8p-deletions occur in 60–80% of primary carcinomas of the prostate, which limits its practical usefulness. Less frequently than chromosome 8-alterations are chromosome 10q-deletions encountered in carcinoma of the prostate. However, the 10q-deletion is of interest, since MXI 1, a suggested tumor suppressor gene, is located on 10q. Another candidate tumor suppressor gene is the E-cadherine gene on chromosome 16q. Since E-cadherine gene codes for a cell surface glycoprotein adhesion molecule, 16q-deletions detected with cytogenetic techniques would be of clinical importance to indicate a higher tendency for metastatic spread through a decreased cellular cohesiveness. Unfortunately, this information cannot be used as a prognostic factor to differentiate between patients who are at risk and those who will not develop metastases.

In essence, prostate cancer cytogenetics is difficult; the molecular cytogenetic methods FISH and CGH require more extensive testing before they can be employed in everyday practice. It may be anticipated that after the investigation of biopsy material or the whole gland questions concerning the choice of treatment, e. g., surveillance or adjuvant, can be answered. A concise basic text on the potential of cytogenetics has been written by Sandberg [8].

The regulation of prostatic growth is covered by *Griffiths* (see his contribution in this book). This chapter provides a key to the understanding of the complicated prostate growth and provides an insight, how therapeutic modalities interfere. Apparently, the epithelial lining of the prostatic ducts changes from the distant to the region adjacent to the urethra. The distant or intermediate regions, for instance, are not subjected to programmed cell death (= apoptosis). However, they contribute to the prostatic secretion. The segment entering the male urethra is responsible for the transport of the secretion as evidenced by the presence of periurethral muscle layers. Moreover, it is subject to apoptosis. In addition, its cuboidal epithelia express exclusively sulphated glycoprotein-2, which is the product of an androgen-suppressed gene similar to the TRPM-2 = testosterone repressed prostate message-2 or clusterin.

Apoptosis is important for the understanding of the therapeutic concept called intermittent androgen deprivation (*Gleave et al.,* this book). This regional variation of morphology and function of the prostate suggests a corresponding variability of response to 5-alpha-dihydrotestosterone to KGF (keratinocyte growth factor) or TGF-β (transforming growth factor β). It fits into the picture that TGF-β, for long was considered to be directly associated with apoptosis, is located in the proximal segments of the prostatic ducts. Here TGF-β and TRPM-2 are implicated in apoptosis. It complicates the picture of how androgen deprivation affects the prostate if one realizes that estrogens may be capable in restraining apoptosis. It is now feasable to measure in prostatic tissue the variety of growth factors. However, it is unclear if this will be helpful in treatment decisions.

On the contrary, the finding of neuroendocrine cells in the prostate distributed throughout the acinar tissue may be linked to the response to hormone therapy. However, in advanced carcinomas, hormonal therapy is practically the only choice. Thus the presence of neuroendocrine cells in biopsy specimen does not really help the patient (*Heicappell,* this book).

A new concept of dietary prevention of carcinoma of the prostate is based on epidemiological, biochemical, as well as on molecular biological studies. The degradation products from isoflavonoid-containing soya and lignan-containing linseed, which are also called phytochemicals, act as weak estrogens and seem to compete with testosterone and endogenous estrogens. Apart from being antiestrogenic – since they may restrain the development of female breast cancer – they have also been called "natural tamoxifen".

Oncogenes and prostate cancer

For the development of a cancer, the activation of oncogene and loss of tumor suppressor genes are apparently essential (*Peehl,* this book). It is estimated that of up to 100.000 genes in our genome approximately 50 can act as oncogenes [10]. Each oncogene was named after the virus in which it was first discovered, thus the name myc stands for "avian myelocytomatosis", ras stands for "red sarcoma virus", and erb-B stands for "avian erythroblastosis virus". The function of proto-oncogenes and derived oncogenes is explained by Weinberg [10] in the following way: each cell possesses a complex machinery that enables it to receive these signals, process them, and launch a growth program. The machinery consists of an array of proteins that functions to acquire growth-activating signals and transmit them throughout the cell. These proteins include cell surface receptors that recognize the presence of growth factors in the extracellular space and transmit signals into the cell's interior, informing the cell of these encounters; cytoplasmic signal transducers that become activated by these receptors and then pass signals further into the cell; and nuclear transcription factors energised by the cytoplasmic signal transducers and in return respond by activating whole banks of cellular genes. The activated gene banks together orchestrate the cell growth program, thus the cell can grow and divide.

Proto-oncogenes encode many of the proteins in this complex, signaling circuitry that enables a normal cell to respond to exogenous growth. Oncogenes participate in this signaling circuitry by specifying apparently functioning versions of the components of this circuitry. Oncogene proteins succeed in activating these signaling circuits even in the absence of stimulaton by extracellular growth factors. The practical importance of the most intensively-studied oncogenes in prostate cancer: ras, c-erbB-2 and PTI-1, is not quite clear. Their expression may have prognostic significance, since molecular changes accompany the development from normal prostate to androgen independent cancer [7]. This is demonstrated in figure 1. Strategies to block the aforementioned

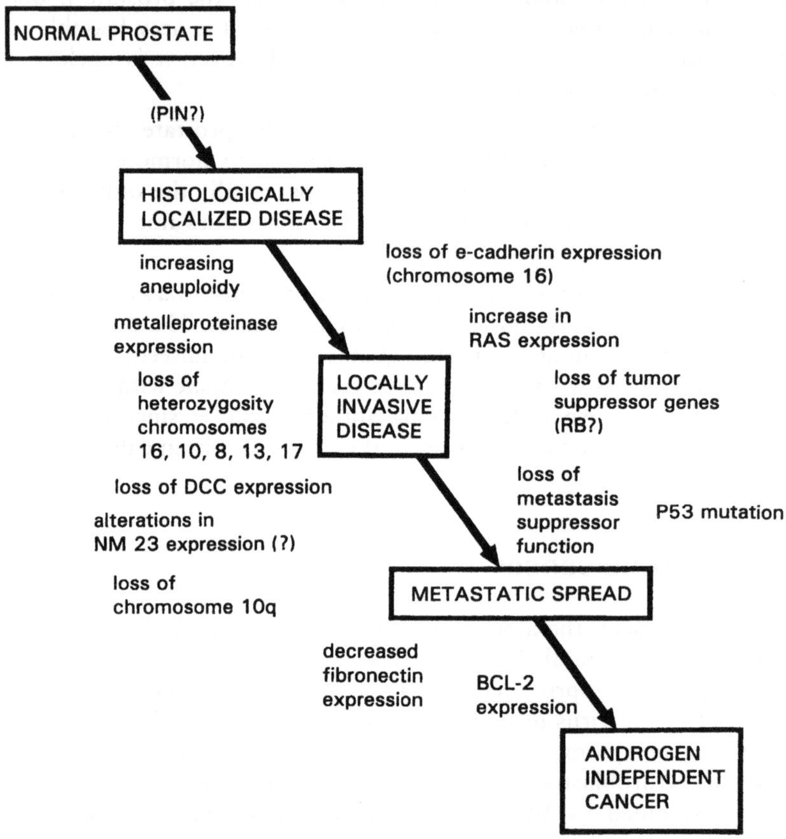

Fig. 1 Initiation, promotion, and progression of carcinoma of the prostate and discovered molecular changes as depicted by Rinker-Schaeffer et al. [7].

growth factors include the use of retinoic acid, which inhibits the prostate but not the tumorgenic cells which have lost their ability to respond to retinoic acid. This may explain why the response to vitamine A1-acid falls behind expectations.

Tumor suppressor genes and prostate cancer

The urologist may be quite familiar with the p53 tumor suppressor gene, which is unfortunately altered in many cancers. Furthermore, p53 appears to act late in the disease, thus it cannot be used to predict the prognosis in early carcinoma of the prostate. *Brewster* (this book) in his account shows that there are at least 20 candidate prostate tumor suppressor genes on 13 chromosomes. The following 5 tumor metastases suppressor genes are apparently identified: KAI 1, p53, E-cadherine, Rb-1 (retinoblastoma), and C-CAM (cell adhesion molecule). Rb-1 and C-CAM, which is already expressed in high grade PIN, act early in prostatic tumorigenesis. However, the methods to detect allelic losses and tumor suppressor gene mutation are continuously improved: Polymerase chain reaction (PCR), in situ hybridisation or Differential Display PCR (*Pilarsky*, this book) are far too time-consuming and expensive to be included in the work-up program of a prostate cancer patient. Furthermore, the results from various studies are still at variance to justify their inclusion in the daily practice. The situation is somewhat different for p53 and bcl-2 oncogene (= B-cell lymphoma), since immunohistochemical antibodies became available. Both genes were found to be independent prognosticators when measured in radical prostatectomy specimen [2]. An interesting prospect is the applicability of Differential Display PCR for prostate biopsy samples or the utilisation of C-CAM as a candidate for prostate cancer gene therapy.

Impairment of cell-adhesion

Among the cell-adhesion molecules the cadherin-catenin-complex is important and its defect may reflect a poor prognosis (*Bussemakers et al.* and *Paul et al.*, this book). Since the cadherin-catenin-complex can be measured by histoimmunochemical staining this allows for potential wide use in practical urology [5]. The main impediment to include the expression for the application as a diagnostic tool is the multifocality and heterogenity of prostate cancer. The intrinsic sampling artifact – only a third of the biopsy specimen have the same Gleason-score as the surgical specimen – will limit is usefulness. Other substances affecting cell adhesion and proliferation such as the proteoglycans (*Schamhart et al.*, this book) are not studied extensively, but it is thought they may act as modulators in cell-cell or cell-matrix interaction. Unfortunately, no practical use is possible, since the well-known SP54 interferes. Along these lines, expression of the androgen receptor during all stages of prostate cancer development will be of practical importance in the future (*Klocker et al.*, this book). At the present time, however, we do not have the means to interfere in the androgen signaling cascade. It should be kept in mind that antiandrogens activate the mutant androgen receptor and that mutations of the androgen receptor gene (= transscription factor) are four times more common in hormone-resistant than hormone-naive prostate cancers. This is one underlying cause for the antiandrogen withdrawal syndrome, which is a drawback of maximal androgen-blockade.

Antiandrogens

These interfere with the action of androgens (= end organ-antagonists) (*Furr*, this book). The receptor-binding of bicalutamide (Casodex®) to the wild type androgen receptor is 2–4 times more potent than flutamid (many generics). If antiandrogen insensitivity occurs – eventually leading to clinical relaps, apart from androgen receptor mutation amplification can be responsible. The latter phenomenon has been found in 28% of hormone-resistant prostate cancer leading to an increased number of androgen receptors. At this stage the tumor may require a higher dose of antiandrogens. Whether micronised Casodex® with its increased bioavailability is advantageous remains to be seen. Finally, mutated receptors might be stimulated instead of blocked by antiandrogens. Casodex® apparently still behaves as a pure antiandrogen and does not activate the mutated androgen receptor.

Drug resistance and new therapeutic concepts

The multidrug resistant gene, MDR 1, and its substrate P-glycoprotein are expressed frequently not only in prostatic carcinoma, but also in the normal prostate (*Siegsmund et al.*, and *Licht*, this book) This puts in question the carcinome specificity. However, P-glycoprotein-inhibitors might be beneficial to reverse chemoresistance.

A sophisticated novel attempt to tackle advanced, even metastasized prostate cancer is gene therapy (*Chung et al.*, this book) (fig. 2; background information: [6]).

Models have been developed which allow for the utilization of proliferation blockers like p53, p16, or p21 – tumor suppressor genes or the toxic gene osteocalcin thymi-

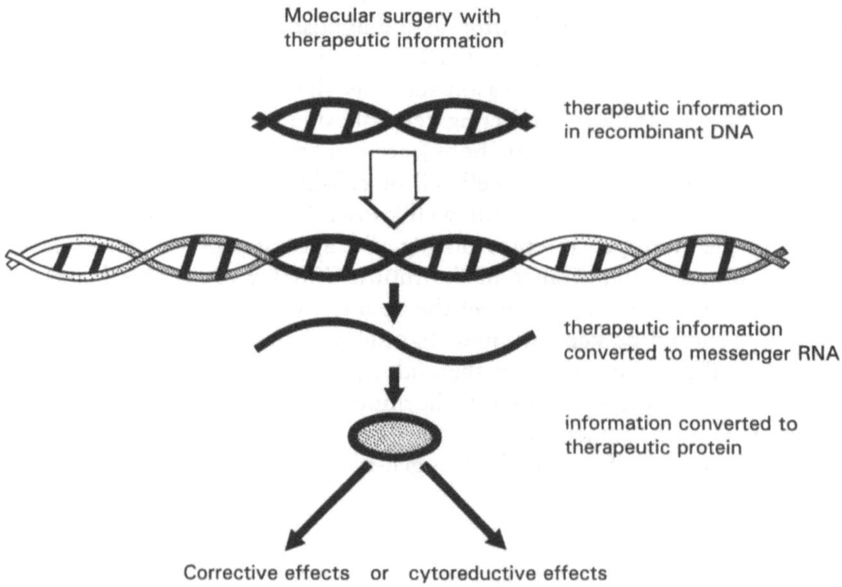

Fig. 2 In gene therapy the information in the DNA molecule is converted into RNA and protein to program either genetic correction or destruction of cancer cells. Adapted from J. W. Simons (AUA News, 1997).

dinkinase which are delivered using adenovirus and a PSA promotor for better specificity (tab. 1).

Table 1 Approved human gene therapy trials in prostate cancer

Primary Investigator	Vector	Gene	Ex Vivo	In Vivo
Chen	Vaccina	PSA		Yes
Scardino	Adenovirus	Thymidine kinase		Yes
Paulson	Retroviral	IL-2	Yes	
Gansbacher	Retroviral	IL-2+gamma IFN	Yes	
Steiner	Retroviral	Anti-c-myc		Intratumoral
Simons	Retroviral	GM-CSF	Yes	

TMC Worldwide Gene Therapy Enrollment Report (June 1996)

Another novel concept to prostatic carcinoma is the anti-angiogenic strategy (*Marmé*, this book). Angiogenesis is related to tumor growth and counting the density of microvessels is a diagnostic method. The tumor vessels grow when stimulated by the vascular endothelial growth factor (VEGF), which is expressed by a variety of mechanisms, e. g., ras oncogene activation, p53-suppressor gene inactivations, or steroid receptor stimulation. The development of an anti-VEGF strategy to "shrink" the tumor has advanced to the point where phase I- and II-trials are scheduled.

Similarly, antimetastatic approaches are under investigation (*Heicappell*, this book). Unfortunately, none matured to a level that a study would be feasable. It is sobering to learn that even the bisphosphonates (Pamidronic acid, Clodronic acid, etc.) do not possess specific activity on the growth of bone metastases from prostate cancer.

The situation is better with the concept of *intermittent androgen deprivation (IAD)* (*Gleave et al.*, this book). Androgen deprivation for metastatic carcinoma of the prostate is usually palliative, thus, following surgical castration, 50–60% of patients progress after 16 months. This may be due to clonal selection, up-regulation of androgen-repressed genes and subsequent abortions of apoptosis of adapted cell subpopulations, androgen receptor mutation, and other mechanism (*Bussemakers et al.*, this book). The IAD should extent the response to androgen deprivation in metastatic carcinoma. Presently, four phase III-trials are open for recruitment under the auspices of SWOG, EORTC, and AUO (2 trials).

Molecular diagnosis of prostate cancer

Undoubtedly, PSA is the best tumor marker. Reverse transcriptase PCR (RT-PCR) for PSA offers a new insight into the work-up and follow-up of patients (*Ellis et al.*, this book). Whereas a second molecular diagnostic tool, RT-PCR for prostate specific membrane antigen (PSMA) appears to be less specific with undetermined biological significance (*Schmidt* et al., this book). PSA RT-PCR may be used to detect "preclinically" pelvic lymph nodes from prostate cancer; this could help to distinguish between patients requiring immediate versus delayed androgen deprivation. PSA RT-PCR positive nodes, although histologically negative, would be an argument against adjuvant irradiation after radical prostatectomy because of an unfavorable histology or positive resection margins. Ultimately, RT-PCR PSA may aid in staging and selection of patients for the appropriate therapy, but the main advantage of this diagnostic tool is its ability to give a molecular prognosis (*Olsson et al.*, this book).

References

[1] Altwein, J. E.: Immuntherapie des Prostatakarzinoms. In: Rübben, H., M. Goepel, B. J. Schmitz-Dräger (eds.): Immuntherapie in der Uroonkologie, pp. 169–179. Springer, Berlin 1993.

[2] Bauer, J. J., R. R. Comully, I. A. Sesterhenn et al.: Biostartical modelling using traditional variables and genetic biomarkers for predicting the risk of prostatic carcinoma recurrence after radical prostatectomy. Cancer *79* (1997) 352–362.

[3] Fauci, A. S., S. A. Rosenberg, S. A. Sherwin et al.: Immunmodulators in clinical medicine. Am. Intern. Med. *106* (1987) 421–433.

[4] Kendrew, J., E. Lawrence: The Encyclopedia of Molecular Biology. Blackwell Science, Oxford (U. K.) 1994.

[5] Martin, S.: Lösliche Formen von Adhäsionsmolekülen. Dtsch. Ärztebl. *92* (1995) B-2172–B-2176.

[6] Nikol, S., B. Höfling: Aktueller Stand der Gentherapie. Dtsch. Ärztebl. *93* (1996) B-2050–B-2058.

[7] Rinker-Schaeffer, C. W., A. W. Partin, W. B. Isaacs et al.: Molecular and allular changes associated with the acquisition of metastatic ability by prostate cancer cells. Prostate *25* (1994) 249–265.

[8] Sandberg, A. A.: Cancer cancer cytogenetics for clinicians. CA Cancer J. Clin. *44* (1994) 136–159.

[9] Wagener, C.: Einführung in die Molekulare Onkologie. Thieme, Stuttgart 1996.

[10] Weinberg, R. A.: Oncogenes and tumor suppressor genes. CA Cancer J. Clin. *44* (1994) 160–170.

Bei Fragen zur Produktsicherheit wenden Sie sich bitte an:
If you have any questions regarding product safety,
please contact:

Walter de Gruyter GmbH
Genthiner Straße 13
10785 Berlin
productsafety@degruyterbrill.com